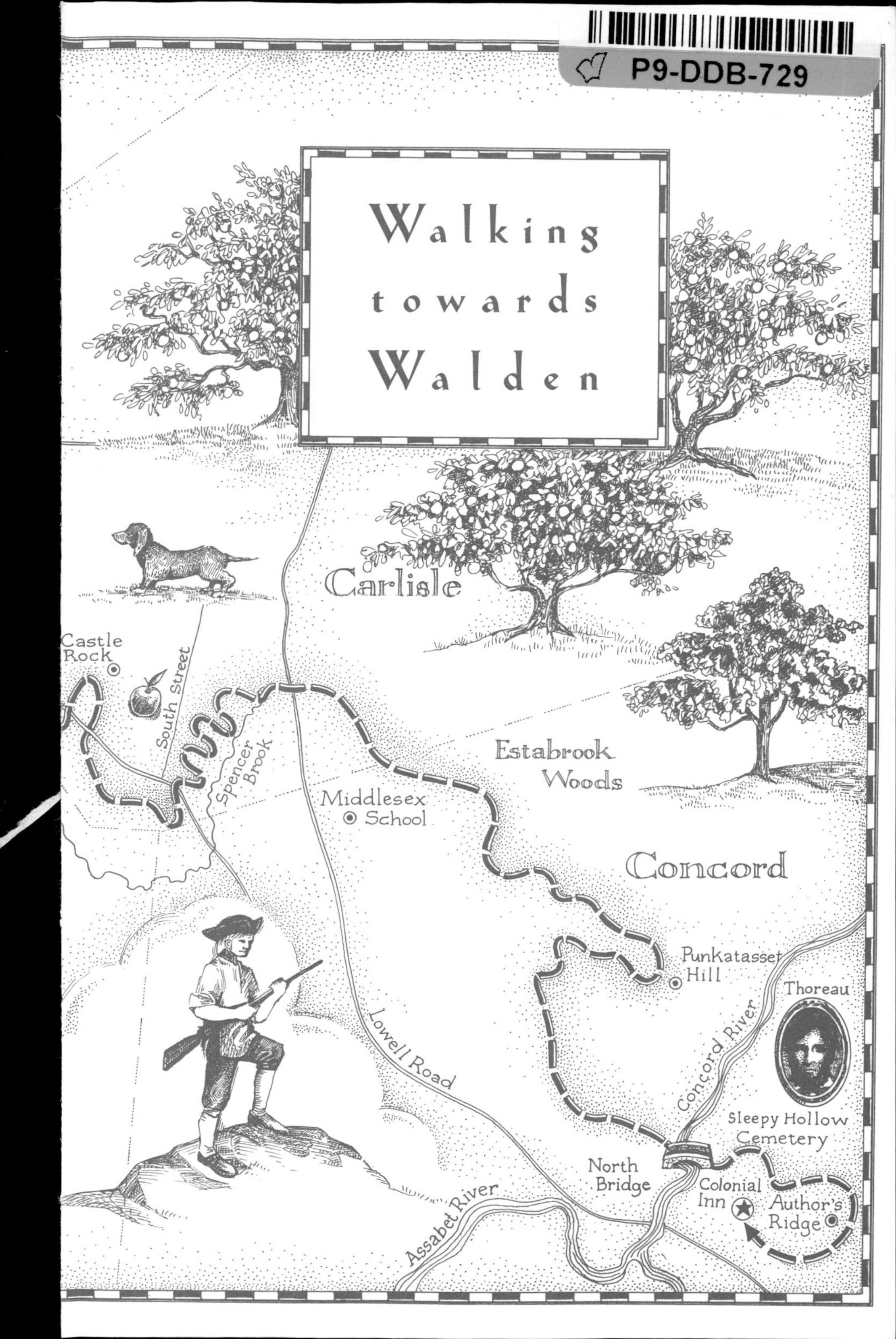
P9-DDB-729
Walking
towards
Walden
Carlisle
Castle Rock
South Street
Spencer Brook
Estabrook Woods
Middlesex School
Concord
Punkatasset Hill
Thoreau
Lowell Road
Concord River
Sleepy Hollow Cemetery
North Bridge
Colonial Inn
Author's Ridge
Assabet River

Also by John Hanson Mitchell

The Curious Naturalist

Ceremonial Time: Fifteen Thousand Years on One Square Mile

A Field Guide to Your Own Back Yard

Living at the End of Time

Drawings by Robert Leverett

A Merloyd Lawrence Book

Addison-Wesley Publishing Company

Reading, Massachusetts Menlo Park, California
New York Don Mills, Ontario
Wokingham, England Amsterdam Bonn
Sydney Singapore Tokyo Madrid San Juan
Paris Seoul Milan Mexico City Taipei

Walking towards Walden

A Pilgrimage in Search of Place

John Hanson Mitchell

Many of the designations used by manufacturers and sellers to distinguish their products are claimed as trademarks. Where those designations appear in this book and Addison-Wesley was aware of a trademark claim, the designations have been printed in initial capital letters.

Library of Congress Cataloging-in-Publication Data

Mitchell, John Hanson.
Walking towards Walden : a pilgrimage in search of place / John Hanson Mitchell.
p. cm.
"A Merloyd Lawrence book."
ISBN 0-201-40672-1
1. Concord Region (Mass. : Town)—Description and travel. 2. Walden Woods (Mass.)—Description and travel. 3. Natural history—Massachusetts—Concord Region (Town) 4. Natural history—Massachusetts—Walden Woods. 5. Walking—Massachusetts—Concord Region (Town) 6. Walking—Massachusetts—Walden Woods. 7. Thoreau, Henry David, 1817–1862—Homes and haunts—Massachusetts—Walden Woods. I. Title.
F74.C8M58 1995
917.44'4—dc20 95-20789
CIP

Jacket design by Sara Eisenman
Text design by Greta D. Sibley
Map by Mary Newell DePalma
Drawings by Robert Leverett
Set in 12-point Bembo by Greta D. Sibley

1 2 3 4 5 6 7 8 9-CRW-9998979695
First printing, September 1995

To
Giorgio
and
Carolina

Let us make pilgrimages to Concord ...

—Last written words of John Burroughs

Contents

	Place	1
	Pilgrimage	9
MILE ONE:	Prospect Hill	11
MILE TWO:	Providence Hill	29
MILE THREE:	Vine Brook	47
MILE FOUR:	Nonset Brook	69
MILE FIVE:	Nashoba Brook	91
MILE SIX:	North Street	105
MILE SEVEN:	Nagog	117
MILE EIGHT:	Castle Rock	139
MILE NINE:	Spencer Brook Marsh	157
MILE TEN:	Thoreau Country	175
MILE ELEVEN:	Estabrook	195
MILE TWELVE:	Mink Pond	213
MILE THIRTEEN:	Punkatasset	229
MILE FOURTEEN:	The Bridge	253
MILE FIFTEEN:	Shambhala	273
	Epilogue	299
	About the Author	301

...the land they purchased of the Indians, and with much difficulties traveling through unknowne woods, and through watery scrampes, they discover the fitnesse of the place....

—Edward Johnson, *Wonder Working Providence,* 1636

Place

THERE IS A SCENE IN THE 1940 FILM *Northwest Passage* in which a member of the battle-weary, starving company called Rogers' Rangers goes crazy in the midst of the Canadian wilderness and announces that he is going home to have his supper. The men stare at him incredulously and ask where he intends to find his dinner in this godless wasteland. He gazes back at them with glazed eyes. "Concord," he says, "I'm going home to Concord to have my supper." Then he turns and races off into the trackless wild.

More than any other community in America, Concord, Massachusetts, evokes what the Hopi people call *túwanasaapi,* the centering place, the place where you belong, the spiritual core of the universe. The town is among those fast-disappearing communities that actually have real main streets with shops and local gossip and dogs that lie down in the middle of the sidewalks in the midday sun. Because of an accident of history, or geography, or some mystical, as yet unidentified force, things have happened in the place that is now called Concord. For five thousand years the local Native Americans would congregate in the area at the confluence of the Assabet and Sudbury rivers, which is located more or less in the present geographical center of the town. The tract was the first inland community to be settled by Europeans. It was the place where the world's first war of independence began; it was the place where American literature first flowered; and it is a place that to this day attracts writers of one species or another—some four hundred have lived or worked here in the town's short history. It is also a place where people continue to live, a place in which the American experiment, with all its energy, its wealth, its economic problems, its violence and perversities, is still developing.

Some time ago, in order to entertain two friends of mine who were visiting from out of state, I proposed that the three of us pay homage to Concord and its luminaries by walking there from a burial site in Westford, Massachusetts, where, according to local legend, a seafaring Scottish earl named Henry Sinclair, in search of the Holy Grail, carved an armorial in stone some one hundred years before the arrival of Columbus.

There are many roads that converge in Concord, both literal and figurative, but I proposed that we attempt to walk there through a precolonial landscape such as Sir Henry might have known. This would mean that we would have to bushwhack approximately sixteen miles through woodlots, old fields, farms, backyards, swamps, and streambeds, following animal trails and woodland paths and avoiding wherever possible all evidence of colonial and postcolonial settlement. It would mean that we would probably get lost, that we would risk arrest for trespassing, and, since we undertook our pilgrimage in October, that we could be mistaken for birds or deer, and be shot by excited hunters. But no matter. We forged on with our plan.

My two companions on this trip were adventurous sorts. One was a basket maker named Katarina Grant, who had made herself into an authority on Native American basketry and had spent a number of years traveling around the country talking to the last native traditional basket makers of various tribes. The other was an old friend of mine named Barkley Mason, who, having collected a number of advanced degrees in comparative literature, chose to make his living by working in the gardens of rich people, thereby leaving open his winters to travel to distant climes to look for birds. They were both sufficiently eccentric to appreciate the folly of our proposed adventure, and imaginative enough to still believe that myth is a continuing presence in landscape. They were also among the least rooted people I knew, having scoured the world in search of a good place to live before settling—if that is the term—in a small rented house in an orchard in the Berkshire Hills. I had been with them on some of their excursions, most recently an absurd venture to Florida to locate the purported Fountain of Youth of Ponce de León fame. But the destinations on these various expeditions were insignificant. Both Kata and Barkley felt, as I did, that the exploration of true places occurs not by design but by accident. True places, as Melville says, never appear on maps.

There are no dramatic vistas northwest of Concord. This is a country of low wooded hills, of swamps, thick forests, and rain, a region that is characterized by human settlement, by the shades of history and literature. Although the area is apparently rural, the landscape is neither picturesque nor sublime—no stormy seacoast, no vast cataracts or high peaks. This is not a country for vistas; the great painters of American landscape traditions went elsewhere when they needed inspiration. But when they set out to shape the concept of landscape, the deeper meaning of place, they looked to Concord.

What interests me in excursions of this sort is the exoticism of the familiar. I am not drawn by expeditions into remote, untrammeled territories where no man, woman, or child has set foot in a thousand years. What lures me is that undiscovered country of the known world, the extant local landscapes where human events have flourished and faded. What I want to do is decode what the Spanish call *querencia,* the deep and abiding allegiance some people feel for a given piece of land. I want to understand why the singular desire of devout Jews is to be buried in the Holy Land, why Moslems prefer to die at Mecca during the Hadj, and why dying Hindus have themselves transported to the ghats of Benares so as to expire on the banks of the sacred river Ganges. I want to decipher this power of place and understand why some groups have such a singular devotion to a spot that they will fight and die to preserve it, whereas others—many of us in the United States, for example—would just as soon move on and let the world behind us end in destruction. I have it in my mind that the answer to these mysteries is as likely to be found in my own backyard as it is in the world at large. As Henry Thoreau said, it is not necessary to travel around the world in order to count cats in Zanzibar.

In spite of the fact that Concord's main attractions—Walden Pond, the North Bridge, the historic houses—are unassuming by world standards, over the past hundred years the town has developed into an American pilgrimage site. Today one million visitors descend on the community annually. This compares favorably with some of the great religious pilgrimage sites of the world. Mecca gets approximately one million pilgrims during the Hadj; Lourdes attracts some six million a year; every August nearly one million Poles converge on the monastery at Jasna Góra to view the Black Madonna; Gypsies travel to a chapel at Les Saintes-Maries-de-la-Mer in the Camargue to pay homage to Santa Sara; and Buddhist and Shinto pilgrims ascend Fujiyama to pay homage to the sacred summit each year. And, as always, India manages to outdo the rest of the world in pilgrimages: some twelve million souls flock to the banks of the Ganges, only one route among an uncounted number of sacred journeys that take place each year on the subcontinent.

In Concord, this tourist trade, along with an improved local economy, has taken its toll. The town has changed dramatically over the past fifteen or twenty years. The old-timers who used to hang out on the streets are gone. Anderson's Market, the corner grocery store, has disappeared; the old dogs that used to block the sidewalks have died; and the working shops have been replaced by boutiques—you can no longer buy anything "real" in Concord, as one local explained. On the other hand, on any given day there you can hear a multitude of languages and a variety of English and American accents. You may see, among the tourists, some well-known writer or professor from one of the nearby Boston and Cambridge universities; you may see one of the film crews who periodically appear in this place because it exudes New England from every corner. During the school year troops of students from Concord Academy, properly outfitted in nose rings, long skirts, exotic hats, and combat boots, pass to and fro along the streets. In summer, great-girthed families from the American mid-West lumber along the sidewalks feeding on ice cream; delivery trucks catering to the gourmet shops jam the normal traffic, speeding cyclists in gleaming spandex weave in and out between the cars; ladies in Harris tweeds pedal slowly along on black bicycles with wicker baskets. Balloons escape, children cry. Japanese visitors videotape the monuments; great throbbing tour buses disgorge their loads of polyester-clad pilgrims. On any given day in Concord you may see the portly Father O'Brien, in somber black. You may see the local Unitarian minister striding along in khaki and florid shirts, greeting the liberal members of his flock. You may see the last living Arawak Indian, Fox Tree, saunter by, his black locks cascading around his shoulders, his eyes glinting beneath his wide-brimmed hat. Or, as I often do, you may fall into discussion with Sudsy Capone, a veteran, so he says, of the Battle of the Bulge, who, with a band of fellow travelers, sometimes sleeps in an unheated shack in the Estabrook Woods, just north of the town.

And if by chance you happen to go to Concord anytime around April 18—as many thousands do—you may catch sight of grown men and women attired in Revolutionary War uniforms and costumes, marching up and down with fifes and drums and flags in celebration of that day in 1775 when the embattled farmers took a stand at the North Bridge and fired the shot heard round the world.

Concord is America's metaphor for itself, an epicenter, a quintessential place, a vision, a dream, an imaginary landscape in which ideas converge, and then break up and spread into the world beyond. To go there is to be swept into a vortex, to be part of a great circle of time in which past is present and present is future, and Arawaks mingle with Germans and Japanese, with programmers from the local computer companies, with the old rich, and the newly rich, and the disguised rich of the stony New England Brahmin tradition.

And so we three, who are but sojourners compared to the rooted burghers of Concord, set out to walk to this mystic place and in the process bind ourselves to those other pilgrims who have converged on this hallowed ground over the past two hundred years. Call it pilgrimage: a journey to the sacred center, during which we three pilgrims will tell each other stories of remarkable people and far-flung places. Call it odyssey: a heroic quest in which we will pass through a dark forest, endure a series of ordeals, and return to the world bearing a boon for humanity. Call it exploration: a journey into the undiscovered country of the nearby. Call it what you will, our short walk was, for all its absurdity, a splendid wayfaring, undertaken in the spirit of Sir Henry Sinclair, who set out to find the Holy Grail, got lost en route, and ended up on a hill in Westford, Massachusetts, where we began.

Henry Thoreau says in his essay on walking that the word "saunter," which is what we were more or less doing on this trip, is derived from a name applied to the pilgrims of the Middle Ages. He says those on their way to the *sainte terre,* or Holy Land, were called sainte-terrers. Alternatively, and perhaps more fittingly for uprooted Americans such as ourselves, he says it may be that the locals called pilgrims people *sans terre,* that is to say, those without ground or a home, without a place.

Every walk, Thoreau says, is a crusade.

Pilgrimage 

...sometimes passing through Thickets where their hands are forced to make away for their bodies passage, and their feete clambering over crossed Trees, which when they missed they sunke into an unceertaine bottome in water, and wade up to the knees, tumbling sometimes higher and sometimes lower, wearied with this toile, they at end of this meete with a scrorching plaine, yet not so plaine, but that the ragged Bushes scratch their legs fouly, even to wearing their stockings to their bare skin in two or three hours, if they be not otherwise well defended with Bootes, or Buskings, their flesh will be torne: (that some being forced to passe on without further provision) have had the bloude trickle down at every step, and in the time of Summer the Sun casts such a reflecting heate from the sweet Ferne, which scent is very strong so that some herewith have beene very nere fainting, although very able bodies to undergoe much travell, and this not to be indured for one day, but for many, and verily did not the Lord incourage their naturall parts with hopes....

—Edward Johnson, *Wonder Working Providence,* 1636
(description of the approaches to Concord)

Mile One

I will talk of things heavenly, or things earthly; things moral, or things evangelical; things sacred, or things profane; things past, or things to come, things foreign, or things at home….

—John Bunyan, *Pilgrim's Progress*

Prospect Hill 

JUST BEFORE HE SET OUT ON HIS JOURNEY to the netherworld, the great pilgrim Dante Alighieri had to pass through a lion-haunted forest where the straight way was lost. Here in twentieth-century America, there is a gloomy forest of hemlocks just below the summit of Prospect Hill in Westford, Massachusetts. As we descend this fertile slope, the great pilgrim Barkley Mason begins quoting from the *Inferno.* He touches his breast and, with a grand sweep, spreads his right arm toward the dark wood below us. "'Nel mezzo del cammin di nostra vita mi ritrovai nella selva oscura—'" he declaims.

Kata is used to Barkley's posturing; she interrupts to ask me something about a mutual friend, and in this manner, we three enter the dark forest and begin our journey.

Today is October 10, five hundred and two years after Christopher Columbus landed on these shores with a company of men and arms. We are standing on a height rising above the central highlands of the eastern seaboard of North America. To the north and west, hills rise and fall to the distant Monadnocks, which give way to the Berkshire Hills, the Alleghenies, the Great Plains, and the Rockies. East of where we stand the world drops seaward through swamps, streams, and marshes, intercut with wooded ridges. This is the known world, the visible land where we will live or die, but all we can see of it from our prospect are trees—a landscape barred with limbs—a stone wall meandering across the top of the hill, rank clumps of grass, sarsaparilla, low-bush blueberry, the blackened, remnant stems of Canada mayflower, brightly streaked red maples, hickories, green oaks, a stately white pine here and there, and below us, like a bad thought, the dark, impenetrable forest of hemlocks.

Our intent is to descend from the ridge of Prospect and Blakes hills, cross the chain of ponds and swamps that make up the headwaters of Vine Brook; and then follow the high ground between the swamps of Nonset, Butter and Nashoba brooks. East of the low ridge beyond Nashoba we will follow the marshes of Spencer Brook and eventually enter into the forest of the so-called Estabrook Country, a hilly tract of old-growth trees that was generally avoided by the local Indian tribes and was deserted by the European settlers as early as 1830. From Estabrook it is an easy walk over the North Bridge, through Concord, and through backyards to the Author's Ridge in Sleepy Hollow Cemetery, where we will pay homage to the incomparable chronicler of this region, Henry Thoreau. From there, if we still count ourselves among the living, we will repair to the Colonial Inn in the center of town to lift a glass or two of hot rum toddy and reward ourselves with a full meal.

That is our general course, but given the digressive nature of my traveling companions, nothing is certain on this walk. Already, not more than twenty yards into our venture, we are distracted. Somewhere in the hemlocks below us, a great gabbling of crows breaks out, like an angry courtroom. They keep up a constant yammering and then periodically burst into a loud clamor and move through the tangle of dark limbs.

"They've got an owl," Barkley says and walks off deliberately to the southwest, the wrong direction. The cawing increases again, and the black horde rises, turns, and swings our way. Barkley raises his glass and watches as a dull, heavy-bodied bird flies by indifferently, followed by the vanguard of the crow pack.

C. Barkley Mason III is in his heaven at such times. The little blackpoll warblers are on the move today, there is a northwest wind and migrating hawks are spiraling overhead, and everywhere, catbirds and wrens and cardinals and white-throated sparrows are whining and whistling in the shrubbery. Between Barkley's tendency to stop for birds and Kata's tendency to slow to a near stop when she talks, I wonder whether we will complete this expedition by nightfall.

This is in essence the same land Sir Henry Sinclair saw on that calamitous day in 1399 when he and a small party of knights in armor supposedly ascended this American hill for essentially the same reason we have—to get an idea of where he was. Sinclair, the legends say, was lost, having gotten blown westward by freak northeasters, and having ascended, out of curiosity, the Merrimack River, before pushing inland to the region that is now known as Westford.

We too are lost, in a manner of speaking, but we have maps of this land, courtesy of the United States Geological Survey, and may be able to get ourselves out of the mess. The maps show us ridges and roads, swamps, streams, ponds, elevations, contours, and steep slopes. They show us not so much where to go in order to get to Concord, as where not to go, that is, how best to avoid roads and developments. Of course the map is out of date, having been updated back in 1987. The land around here is changing faster than the mapmakers can account for.

One hundred and fifty years ago, standing on this height, we probably could have seen the church spires of Concord. Prospect Hill was sheep pasture, and with the exception of the steeper slopes, the swamps, and a few isolated woodlots, the land was pasture or fields or garden plot all the way to Concord. Now, with the exception of the farmlands of Littleton to the south, the horse pastures of the idle rich of Carlisle and North Concord to the east, and, of course, the larger backyards, it is all woods. The seventeenth century has reasserted itself and we will be under trees for the next fifteen miles.

This is not much of a walk, really, a mere stroll compared to some of the forced marches my companions have put themselves through in the wilder regions of this planet. But then it is autumn, the weather is of the type that is inscribed forever in the American mind as high New England—blue skies and flying clouds—and the forest below us is flaring up in garish autumnal splendor. No place else on earth save a few forested regions of northern China puts on such a brilliant display. But although we are placed here in a new land, in young America, so to speak, we are in fact ever accompanied by history, and myth, and legend. This land is an old land; an ancient, thriving people lived here when Sumer was in her youth. They endured when Greece was in her glory, and were still here when the grandeur that was Rome was overrun by barbarians. For fifteen thousand years, the unchronicled nations of Native Americans created a history for the continent, but because it was unwritten, and because it was their history and not ours, we are told that time begins in Concord in the year 1635.

The open woods of the summit of Prospect Hill is marred by a chain-link fence surrounding a gigantic blue water tower that serves the town of Westford. Just below the summit, across an old stone wall, the sloping ground is forested with oaks and hickories interspersed with a few remnant yews left over from the time some fifty years back when this hill was pasture. Growing among the yews, in this season of decay, are mushrooms—lactarius, collybias, and russulas. At one point in our descent, Barkley stoops to inspect an old stump and comes up with a rare species, a hideous phallus-like thing he identifies as a netted stinkhorn. "Smell," he commands. Kata sniffs and retches dramatically. It has a fetid deathlike odor.

A few hundred yards farther along, we move into the gloom of the hemlocks.

By tradition, the dark forest is dangerous for innocent pilgrims such as ourselves; we enter at risk. In the myths of Western civilization, the forest represents a place beyond the bounds of the known world, a place where pilgrims and hunters get lost, where you may encounter wild beasts, evil dwarfs, witches, gnomes, and snarling trolls. Magical transformations take place here, bears become princes, fairy courts hold torchlit processions by night, wayward children are captured by witches and become toads, damsels disappear for a hundred years until they are restored to life by the kiss of an adventurous knight.

But all true pilgrims, all heroes, have to pass through the forest. According to the legends, there is bounty somewhere in here, or beyond, and the virtuous soul can find the equivalent of the Holy Grail and thus restore a land laid waste by sickness and death. The pilgrim recapitulates the hero's journey, and for us, as adventurers in the spirit realm, as heroic travelers, this walk is all part of our work today. And so we steel ourselves against adversity and enter the woods.

Not a good beginning, though. The hemlock trees are growing close together in this place, the lower limbs are dead and more or less entwined; we have to duck under them and at the same time clamber over fallen trunks overgrown with pincushion, fern, and haircap moss and strewn with tiny brown mushrooms with long stems. Halfway over one fallen log Barkley picks one of the mushrooms and sniffs it.

"Ah-ha," he says. "The rare and endangered LBM."

"What's an LBM?" I ask.

"You know, LBM. Little brown mushrooms of some sort. I haven't a clue what they are. It would take me a year to identify these, but I thought for a minute they were out-of-place fairy rings."

Fairy ring mushrooms, which often grow in a circle, spring up in the early morning, according to legend, in those places where the forest-dwelling fairies danced the night before.

A snag, a treacherous gnome's finger, catches Barkley's pack and then snaps off. He has to stop and untangle the strap and in the process tears it.

For all his freewheeling travels, Barkley is a prisoner of catalogues. He scours them each fall and spends half his money on outdoor equipment. Like all good pilgrims (who traditionally dressed in wide-brimmed hats), he is well clad for our expedition today in an expansive felt outback hat and a Norwegian wool pullover. He has a leak-proof pack with zippered mesh wing pockets, a stainless-steel thermos, a compass, and spanking new trekker boots, which he is hoping to break in on this very walk. His binoculars, which he carries strapped to his chest like the scallop shells worn by medieval pilgrims on their way to Santiago de Compostela, are Swarovskis, the newest vogue in birder's glasses.

Kata, by contrast, favors jeans, old, much-sewn Austrian hiking boots, and a heavy wool sweater, which she herself knitted in her pre–basket weaving days. She is wearing her hair in a single braid and has knotted a blue jay feather in the plait. Kata is partial to Native American customs. She has high cheekbones, blond hair, and green eyes and likes to entertain people by telling them she is one of the daughters of Copper Woman, the mythic red-haired mother of a race of Native American feminists of the Northwest tribes. She is well versed in Indian lore and makes her living by teaching "native life-ways" to schoolchildren and selling her natural baskets, some of which are so finely crafted that they have made their way into museums.

At the base of Prospect Hill we come to an ancient stone bridge with tightly joined abutments and two immense stone slabs arching over a water course, which at this time of year is dry. The rain of forest debris has covered the slabs and old road with needles and leaves; small hemlocks and white pines are growing out of the accumulated detritus on the bridge, and the stone abutments are covered in green mosses. Lower down the slope we come to a series of channels and sluiceways, also abandoned and overgrown; nearby is an old pond, marshy and clogged with milfoil and royal ferns.

All this was part of a waterworks that once supplied the town of Westford and was abandoned sometime in the 1950s in favor of more advanced delivery systems. North of us, surrounded by a fallen wrought-iron fence and elaborately sheathed in scalloped, moss-covered cedar shingles, we can see the collapsed pump house, its ancient machinery lurking in the dark shadows. Mice, bats, and digging mammals have taken over the foundation and eaves, and not twenty feet from the iron fence we find the half-rotted body of a raccoon. Twenty-five yards to the south of the pump house is an overgrown stone wall with a vine-covered, fallen gate stuck forever at a cant. Beyond that is a verge of poison ivy, blackberry, and beer cans, and beyond that a rural road known as Hildreth Street.

The abandoned waterworks and overgrown hillside have the nostalgia of decadence and decay, a failed economy, farms worn out, woolen mills deserted, families in poverty and flight, a rich aura of time and history. And yet even as we stand here admiring the exquisite decline, the speedy little SAABs and Audis of a few errant commuters are whipping by. We cannot forget that we are hiking along the outer boundaries of one of the edge cities of America, one of the highway centers that make up what social critics call the new frontier of this continent. The term "edge city," coined by the writer Joel Garreau, describes the new urban centers that have grown up around freeway and airport intersections and are characterized by agglomerations of corporate headquarters, shopping plazas, and, close at hand, condominiums and single-family homes, the whole of it strung together by pathways for the automobile. In the Boston area, mini-edge cities have sprouted at the interstate highway junctions along Route 128, the belt that advertises itself as "America's Technology Highway."

Edge cities do come to an end and at a certain point, beyond their boundaries, the executives of the corporations live. Here, in contrast to the high-rise, high-speed character of the edge city centers, you will find expansive green lawns, woods, open space, country clubs, and horses—the great forest stretching from Concord up to Westford, for example, is interlaced with horse trails.

Some ten thousand people live within ten miles of the old bridge on the lower slopes of Prospect Hill, and yet I daresay not three or four know that the pump house is here with its once fashionable wrought-iron gate, its nineteenth-century machinery, and its sinister undertones. In fact I would venture to say that not more than five or ten even know where Prospect Hill is. By contrast, those who once lived here knew their territory; the maps and old histories of the towns of Westford, Carlisle, Concord, and Littleton are veritable litanies of place names.

Just before we were dropped at the base of Prospect Hill by our loyal shuttle driver (who will pick us up this evening at the inn), we paid homage to the fallen knight in armor memorialized in stone by Sir Henry Sinclair. The image is carved or chiseled in a flat granite slab, much eroded and scribed by glacial scratches and ten centuries of ice and rain. In recent years, as a result of an interest on the part of transnational Scottish clans, much has been written about Sir Henry, and the site has been visited by pilgrims from both sides of the Atlantic. I suspect some of the visitors are disappointed. It is difficult to see the image of a knight in the confusion of scratches and ruts that are a part of the rock's natural face. On the other hand, precisely because of the many scratches and ruts, one can read there any number of images, and when the light is at the proper angle, in the early morning and late afternoon in autumn, for example, you can make out certain aspects.

Sometimes local keepers of this sacred relic emphasize the image by chalking in the punch holes that Sir Henry's armorer made with his chisel, and at these times, conveniently, you can see a full fair helmeted knight, his emblazoned shield held at an angle and inscribed with the mullet, buckle, crescent, and ship that are the emblems of the Sinclair knights. You can also see the wheel and pommel sword that according to the Cambridge University archaeologist T. C. Lethbridge were the weapons of choice of Scottish knights at the time. In fact, it was the sword that identified the knight as a member of the expedition of Sir Henry.

The image on the rock was known to residents of Westford ever since the town was settled, but the sword was thought to be a tomahawk and the carving the work of local Indians. Frank Glynn, an amateur archaeologist from Connecticut, was shown the image in 1946 and thought it looked like a sword, so he made a tracing and sent it to Lethbridge, who wrote back to say that swords of this sort often appear at burial sites along with full images of knights. Glynn went back to Westford and in the proper light managed to see the figure, along with the shield and even the heraldic emblems.

The whole series of Sinclair legends is perhaps questionable; in fact, as far as I know professional archaeologists have not even bothered to refute the story. But this Prince Henry Sinclair, First Earl of the Orkneys, was a historical figure and was indeed a great adventurer and sea voyager. He was born around 1345 at Rosslyn Castle, outside Edinburgh, died in 1400 and is buried in a crypt at Rosslyn Chapel. He came out of a family with association with the Knights Templar, the order that was originally established to protect pilgrims en route to the Holy Land. The cult, originally a selfless, devotional order, amassed great fortunes and power in the Middle Ages, and eventually became a threat to existing kingdoms and duchies. The Templars were disbanded in 1312 by order of the Council of Vienne, their leaders were put to the stake, and the lesser knights were imprisoned or banished. But according to some theories, the Templars simply went underground and reestablished themselves as a secret cult. One of their underground centers was thought to be Rosslyn Castle, which today is connected with ancient Masonic rites.

After the Templars lost their base in Jerusalem and had their wealth confiscated, they began a quest to found a new empire somewhere beyond the reach of the European potentates who had brought about their downfall. The storied land to the west, beyond the Atlantic, was a potential site. According to Andrew Sinclair, a descendant of Sir Henry who wrote a book about the subject called *The Sword and the Grail,* Sir Henry set out for the west to establish a new Jerusalem and build a new Temple of Solomon for the Knights Templar.

Sir Henry set sail from the Orkneys in 1398 with Antonio Zeno, a Venetian sailor and explorer who with his brother had strayed north from the Mediterranean in 1390. The Zeno brothers discovered a series of new lands far to the northwest of Europe, including Estotiland, which later interpreters of the story have identified as Nova Scotia. On their way back, the two explorers were blown off course and shipwrecked in the Faroes, where they were attacked by the natives. While they were under attack a knight appeared with a company of chiefs and rescued them. It was none other than Sir Henry Sinclair, who just happened to be out on an expedition to the Faroes to add the islands to his northern holdings. Sinclair, who is identified as Zichmni in the early manuscript known as the *Zeno Narrative* that recounts this adventure, had a castle in Scotland and some twelve ships, and was anxious to be off for parts unknown. After some negotiations, the Zeno brothers, Sir Henry, and a few members of the Gunn clan set off from the Orkneys. Following the general route taken by the Vikings, they coasted along the shores of Iceland, Greenland, and the reputed colony at Vinland in northern Newfoundland, until they came to a harbor with a smoking mountain and a pit of boiling tar, which some interpreters of the story identify as Pictou Harbor in Nova Scotia.

From this point onward the story gets confused. Antonio Zeno takes the ship back to Europe. Prince Henry remains in Nova Scotia, where he befriends the local people and even works his way into their mythologies—some authorities cite him as the model for the Micmac Indian hero Glooscap. During his sojourn in the west he explored the coasts of the western Atlantic with a company of knights, was blown southward at one point, and ascended the Merrimack River, Stony Brook, and ultimately, Prospect Hill. The legends say that on their way up or down the hill, one of the knights had a seizure and died. Prince Henry and company buried him on the site and carved the armorial, complete with Sinclair emblems, on a nearby flat rock.

The source for this much of the story of the voyage is the *Zeno Narrative* which the maritime historian Samuel Eliot Morison considers to be part of the imaginative literature of the sixteenth century. But documents of this sort were not uncommon in Europe. Medieval legends of distant kingdoms, in particular the vast Asian empire of the benign Christian king, Prester John, charged the imagination and fostered any number of actual voyages, including those of the Portuguese Prince Henry the Navigator, who sponsored many voyages of exploration.

Henry Sinclair never managed to found his new Jerusalem, nor did he rebuild the Temple of Solomon, nor did he find the Holy Grail, which, according to his chronicler, he was also seeking on his westward voyages. He ended his days back at Rosslyn Castle; not one year after his return he was killed in battle while defending his native lands against the invading English.

He lies buried at Rosslyn Castle, surrounded by imagery from the more ancient traditions from which he and his Scottish clansmen had recently emerged. In the chapel at the castle, icons of the old pagan nature cults—

Dante's dreaded dark forest—managed to reassert themselves in Christian iconology. A great treelike pillar in the castle chapel is surrounded at the base by eight octagonal serpents with their tails in their mouths. The tree is reminiscent of the old Norse ash, Yggdrasil, the tree of life, which holds the heavens above the earth and was entwined by a dragon or serpent. Similar images were once part of the ornament in the Temple of Solomon, and appear also in the Great Mosque in Mecca, in the stained glass of Gothic cathedrals, and in Florentine depictions of the life of Christ.

Kata says that in Native American cosmologies landscape is often the very source of creation. Caves, mountains, streams, and trees are sacred, revered places. A hole at the bottom of the Grand Canyon, known as the Sipapu, is the site from which the primal Anasazi emerged from the underworld. An obscure cave on Mount Baboquivari in Arizona is the home of I'itoi, the elder brother, the spirit who created the Tohono O'odham people. The American continent, a barren, meaningless desert to us blind Europeans, is layered with legend for the people who have lived here for the last twenty thousand years.

But we poor non-natives, stranded here on the alien shores of a savage continent, three thousand miles from our ancestral roots, have had to invent a past for ourselves in order to establish our place. And so we envision exotic voyages, and place knights in armor on the slopes of New England hillsides. It all makes sense: everyone needs a center.

We have now crossed one of the five or six twentieth-century roads we will have to traverse in order to carry out our precolonial pilgrimage. Hildreth Street in Westford is a pleasant rural road lined with a few nineteenth-century farmhouses, open pastures, and newer suburban houses set in woods between the pastures and the slopes of Prospect Hill. We skirt a field and climb through woods along the south slope of a ridge, traipsing through patches of poison ivy, jack-in-the-pulpit with strange, bright red fruits, and blossoms of New England aster. Just below us there is a saltbox farmhouse, built, so we are informed by a historic marker, in 1658, one of the oldest extant houses in the area. The house has an assymetrical gable roof whose longer side slopes back to create a low wall on the northwest side, in which there are a few narrow leaded-glass windows. The long roof, low wall, and tiny windows, the whole configuration, is backed up against the hill and protects the structure from the prevailing northwesterly winter winds. Conversely, the higher, front facade of the house has a row of larger mullioned windows, which allow entry of the winter sun, low slung in the southwest sky. Whoever built the house made conscious use of the landscape to site his saltbox for protection from winter winds and for maximum solar gain.

This is the sort of vernacular architecture that has developed a following among landscape historians, planners, and designers in recent years, a sort of "pattern language," in the words of the architect Christopher

Alexander, in which the people of a given region design their living quarters according to the nature of a place, not according to some abstract ideal. It is a language that has now been lost as a result of commercialism, uniform building codes, engineered house designs, and the general lack of imagination that has flattened the human-made landscape of America.

We are attempting to follow an old road that, according to maps, seems to have been abandoned in 1785. The going is rough, though. In our first attempt to find the road we are faced with an impassable wall of poison ivy, so we walk uphill for a hundred yards into the woods and then turn back downhill to circumvent the obstacle. But as soon as we reach the old road again we hit another barricade—this one of brambles. We stand on the slope eyeing our prospective route.

Beyond this brambled wall lie hills, the ring road known as Route 495, then more thick woods, treed swamps, and beyond, the sucking marshes of the Spencer Brook. Beyond that, the Promised Land.

"Come on then," Barkley says.

Holding his arms above the stickers, he wades into the tangle.

Talk of the white doe flowed like a river tumbling from its source in the clefted rocks....

Often she was seen browsing amid the brown herd of deer that lived there. But she always remained apart, turning her head to the east, sad-eyed and dreaming in the direction of the distant sea. Those who were compelled to hunt her said that their arrows, though well-aimed, fell harmless at her hooves—whereupon she would leap with the west wind, swift as milkweed down, bounding the sand hills, driving the quick curlews and iron-winged cranes up into the cold gray slate-colored sky.

—Gerald Hausman, *Tunkashila: A Mythological Saga of Native America*

Providence Hill 

HENRY THOREAU SAYS IN HIS ESSAY "Walking" that in every town there are a few old discontinued roads that may lead to profit, by which he meant adventure. Thoreau favored cross-lot walking—in other words, trespassing. He abhorred fences, boundaries, and horses—he was his own horse, as he said. He claims that to walk out into wild nature is to join with the prophets, poets, and explorers of old, the great metaphorical wanderers of history—Moses, Chaucer, Columbus, and Americus Vespucius. It is here, he wrote, on the local abandoned roads and in the unfenced woodlots and pastures that you will discover the mythologies that are the truer histories of America.

That may be. But we three are finding that the untrodden paths do not offer contemplative strolls. As soon as we approach the road, Kata's wool sweater becomes entangled and she stops and begins picking out the briars. Barkley, whose neotechnic armor sheds the spears and lances of thorny plants, turns suddenly in mid-tangle, his binoculars half raised, and pleads for silence.

"Listen," he says, "blackpolls. What's this? October tenth? They're right on schedule."

Here and there in the high trees we hear a lisping whistle and we can see forms darting about. We back out of the brambles and stand at the edge of the woods and the fields surrounding the farmhouse. Barkley stares at the treetops for a minute and then, without ceremony, hands the glasses to Kata, who, after much fiddling and searching, fails to see the birds and hands the glasses back. "I saw their flight," she says. "It was enough. It's what birds do."

I take the binoculars and see the small black and white forms darting among the green leaves. This is the time of their seasonal passage, the great nomadic cycle that they follow twice a year. Blackpoll warblers nest in the coniferous forests of the north. Sometime in September—flying by night, resting and feeding by day—they begin a long journey south through New England, the Central States, the American South, and then across the Gulf of Mexico to Central America, where they spend the winter. In early April, with their ancient genetic codes firing off as yet ill-understood messages, they undertake the journey in reverse.

The great migratory passages such as these would tend to make you dismiss the existence of any biological devotion to a particular place, but in fact passerine birds such as the blackpolls return year after year to the same small territories, each bird recognizing a space no larger than a few acres in a vast wildness of thousands of square miles of Canadian forest. Barkley says the same is true in their southern wintering grounds. Having flown south across the continent and the wide seas, individual birds seek out the exact area where they spent the winter in previous years, a phenomenon known in the trade as "winter site fidelity."

The blockade of brambles finally proves insurmountable, so we back out and climb again though a mix of oaks and pines, to a light woods in which other wanderers—hunters, no doubt—have passed. We see their spoor, as Barkley phrases it: an old Rheingold bottle, an aluminum pan, some oilcans, and an old mouse-shredded piece of plastic wrapper. Then, deeper in the woods, we come across a strange construction, something that looks like a cross between a chicken coop and a doghouse. The building has been overtaken by brambles and young birch, one of which has grown up through the foundation and then emerged from a window, seeking light. Growing around the doghouse Barkley finds a profusion of lactarius mushrooms.

We can see above us now the large brick house of a woman named Priscilla Eliot, a sprightly eighty-two-year-old who dresses in tweeds, wears steel-rim glasses, and favors sturdy walking shoes. I had met her earlier when I was out scouting this section of the world and explained our mission to her. She is an avid member of the Massachusetts Audubon Society and other local birding organizations, including a group known as the Rowley Dump Bird Club. Members, most of whom are women in their seventies or eighties, meet periodically to scour the New England landscape for interesting bird species. One of their favorite haunts is the town dump in Rowley, hence the name.

Mrs. Eliot's property is a finely manicured landscape in the 1920s style, with a combination of old copper beech trees, trimmed hemlock hedges, fir trees, and clipped lawns, interspersed with flower and vegetable beds. Just the type of place that reminds me of the old ferryboat suburb in which I grew up.

We keep to the east of her property and move through more tangle of blackberries and poison ivy just at the edge of an apple orchard and soon come upon the back of an old, seemingly abandoned, single-car garage set alone at the edge of the orchard. Barkley peers into a cobwebbed window, cupping his hands against the glare.

"My God," he says. "Stopped time. This is right out of Dickens. Miss Havisham's wedding day."

Inside we can see, in the half-light, dusty scythes and hay rakes, a horse-drawn sleigh, a fine leather-seated carriage, a farm wagon painted up like a Gypsy cart, and a beige 1931 Model A Ford convertible with a louvered bonnet.

As we peer in the window, a stocky man in heavy corduroys and a peaked tweed cap appears around the corner and politely clears his throat.

"Just passing through," Barkley says. "On our way to Concord."

"Out for a walk," I add to make things more clear; no one around here actually *walks* to Concord.

He lifts his cap, revealing a full head of curly gray hair, and introduces himself as Jack.

Actually, Barkley, Kata, and I rarely, if ever, just pass through. Stopping is part of the exploration of place and in a few minutes we are deep in conversation.

Jack is the gardener for Priscilla Eliot and he turns out to be a great raconteur. We learn from him that the whole orchard to the east of us, the land we are about to pass through, is doomed.

"Just last night," Jack explains, "they lost the battle to save this orchard. Developers got the okay to tear the place up and build houses. Won't see many more harvests from here. They're going to take it, every last tree, I'll bet."

We are standing at the edge of an old rutted orchard road that runs straight through the trees and then sinks down below the shoulder of the hill. In the far distance, framed by the trees, exactly beyond the road, I can see the outline of Blakes Hill just to the north of our intended route. It's a fine scene, right out of an old Wallace Nutting New England landscape, or an American impressionist painting. In fact the road must have been laid out by some sharp-eyed farmer to reveal the outline of the hill. But it's a doomed land, as Jack says.

"There's a great white-colored buck deer with a huge set of antlers that lives in this orchard," Jack says. "Hunters can't get him. They come here year after year looking for him. Sometimes manage to get a shot off. But he gets away every time, God bless him. Now what's to become of him?"

Jack tells us that one day last year, just before the winter solstice, Mrs. Eliot and the members of her bird club were sitting in the kitchen weaving greens for Christmas wreaths. Just at dusk one of them looked up and saw the great stag standing in the yard, staring in the lighted window at the circle of women. As they watched, he turned and bolted into the orchard.

"Just disappeared into thin air," Jack says.

It's a good story. Also an old story: crones sit in a circle by the hearth weaving circlets of greens from the sacred tree when a white stag appears in the gloaming out of a doomed land. Throughout the ancient world the crone was associated with witchcraft and knowledge of earth lore. The evergreen bough was sacred, the circle, even more sacred, a symbol of the eternal cycles of life and death, and the hearth was the heart of the home, the core of what the Germans call *Heimat,* a combination of home and place. The stag, by contrast, was considered the protector of wild places, the guardian, and the messenger between the wilderness and civilization. He appears in ancient India in the form of a horned god, a pre-Shiva deity. He is found in Greek mythology, as the companion of Artemis, the goddess of the hunt. He appears in Celtic mythology as Cornnums, the stag man. By the time Christianity arrived in Europe the stag had evolved into the companion of saints and crusaders. A mystic white stag guided Charlemagne through the Alps during his campaign to save Rome from barbarians. By the Middle Ages in Europe, the deer had become the symbol of more Christian saints than any other wild beast. In one popular medieval story, the pagan hunter Eustace came upon a great white stag in the forest and drew back his arrow to slay it, when a cross appeared between the stag's antlers. "I am the Redeemer," the stag said to Eustace. "Whyfore dost thou persecute me?" Eustace fell to his knees and converted on the spot. Later he was made patron saint of hunters.

Kata has spent enough time among indigenous peoples to have learned that the great mysteries of life are often revealed in daily events. Some of these incidents evolve into myth, and the place where they occurred is enshrined. Among the Native Americans she knows, who eat fast food, smoke cigarettes, and watch a lot of television, everyday events create stories that are carried on from generation to generation—an unexplained car wreck, a murder, a ghost sighting, the appearance of a black dog with red eyes, even personal memories, come to be associated with certain sections of the landscape which then become a part of the story. Nothing plays out in an empty room, stories create landscape—or vice versa: the land begets the story. Mecca was a holy spring a thousand years before Mohammed got there. Lourdes was a well-known grotto above a river on the pilgrim's way to Santiago de Compostela before Bernadette saw the Virgin there. Gay Head cliffs on Martha's Vineyard were deposited by the glacier eight thousand years before the Wampanoag Indians stood in awe of the place and came to believe that it was the home of their chief god, Marshope. Geography engenders mythology.

Unfortunately, here in the East, the first European chroniclers of the places that were considered sacred by the Pawtuckets, Nipmucks, and Massachusetts tribes chose not to set down a record of their identity; in fact they did their best to obliterate any evidence of what they considered devil worship and paganism.

The orchard rolls eastward; the land slopes. We hike down the clear dirt road at a good pace now, telling stories of holy mountains and sacred stags.

"A friend of mine in Acoma says he saw a deer leap out of the clouds," Kata says. "It appeared in the same quarter where the Indians saw Santiago come down out of the sky with a fiery sword during the Spanish siege of the town. The Spanish never could understand why the Indians surrendered; they had broken the Spanish siege when suddenly all the Indians fell back in terror. Later they told the Spanish they saw a huge knight emerge from a thunderhead on a white horse. 'That was Santiago,' the Spanish explained. 'We didn't even know he was fighting with us that day.' "

"Well what about poor Actaeon?" Barkley says. "All he did was happen upon Artemis while she was bathing. She looks around for her spear and since she can't reach it, splashes water on him, and the next thing, he's got these horns. Then his own dogs chase him and he can't even call them off because now he's a full-blown stag, so the dogs down him, and along comes one of his fellow hunters and delivers the coup de grace."

"Mortals shouldn't try to look at God," Kata says.

I tell them about my brother's adventure with a fawn. He and I were hiking in the Shell Canyon in Wyoming, and at one point he went on by himself, up to a clearing in a small green valley. He was resting there when he heard a scrambling in the brush, and a fawn, drenched in perspiration, dashed out of the cover, ran up to him and stood next to him, its body touching his legs. Before he could even react, a huge "wolf" (no doubt a big coyote) charged out of the brush and, without breaking stride, ran past them and disappeared.

"Ah-ha," Kata says, "the Pawnee would say your brother's got *maxpé,* spirit power. If he believed in any of this he'd take the deer as his totem animal and wear horns. He'd go to the big Pawnee deer dances they hold. Maybe he'd even be able to see the deer woman; they say she still shows up from time to time. Dancers saw her at a big pow-wow back in 1975, a normal-looking woman, but if you look down below her skirts you see that she's got these hoofs. She seduces men and then tramples them. Some guys followed her out of the stadium the last time she showed up and they found their bodies out in the scrub, all bloodied up and trampled."

I like the image of my brother, who favors conservative, boring clothes, dressed in antlers and a loin cloth.

"You think this deer woman would trample my innocent brother? He's a social worker."

"Probably not, if he's got *maxpé*."

Halfway down the hill we begin to hear a vast roaring, as if we are coming upon a waterfall. It's the sound of Route 495, a major north-south highway that encircles Boston, beyond the inner circle of Route 128. These great ring roads are exciting, active places, charged with angry traffic, steel and glass, and attempted landscaping. Barkley sees them as modern versions of Dante's nine circles of Hell, each designed to torture the greedy consumers in its own way. Route 495 is attempting to grow an edge city of its own about twenty-five miles south of here, at the junction with the Massachusetts Turnpike in Westborough. The two highways cut through Cedar Swamp, which was the outpost of one of the last pure-blooded Indian families in this part of the world. On cold winter days in the 1920s, they used to emerge from the swamp and beg milk from the local farmers. The swamp is also the headwaters for the Sudbury and Concord rivers, which were critical players in the American literature of landscape, the metaphorical hunting grounds for Thoreau, Emerson, and Hawthorne. The Sudbury meets the Assabet at Egg Rock in Concord, at the place the local tribes called Nawshatuck, a spot some believe was an important seasonal and ceremonial gathering site.

In our time the Sudbury and Concord are experiencing a spiritual renaissance. Several books about the rivers have been published in recent years; naturalists are taking a renewed interest in the waters; there is a national wildlife refuge strung along the river meadows; the photographer

Frank Gohlke, who had worked mostly in the emptiness of the American West and mid-West, spent four years concentrating on a single stretch of the Sudbury; and because of the efforts of local conservationists with a bioregional slant, Cree Indians from Canada come to the rivers periodically to celebrate the gathering of waters and emphasize the need for protection of the waters of the world.

Barkley, Kata, and I had been debating this Route 495 problem since breakfast—mainly how to cross it. It is the major obstacle we will encounter on our walk: protected in some areas by high chain-link fences, sunk beneath deep barriers of earth in others, surrounded by concrete, and of course running with dangerous, speeding traffic. Even now, beyond the trees of the orchard, we can hear the high pitch of whining semitractor trailers charging north and south to points unknown.

After a few minutes, we break out onto an embankment above the highway and see the racing line of cars and trucks. Rush hour is coming to an end, but the tide is still running full. In fact, like all epic heroes, we have come to the traditional clashing rocks—immense, hard stone gates that roll open and then close on incautious travelers, crushing them. If the adventurous hero can get through the clashing rocks, he or she can pass beyond the restrictions of the visible world and gain access into the spirit realm or, as post-Freudian interpreters suggest, voyage inward to self-exploration and discovery.

Jason and his Argonauts had to pass clashing rocks known as the Symplegades on their way to Colchis to capture the Golden Fleece, which hung in a sacred grove and was guarded by a monster. Jason had been warned to release a dove before he attempted passage, and when the heroes came to the great rocks, they slowed the vessels and let the dove go just as the rocks rolled open. The dove made it through, and the heroes laid back on the oars and managed a safe passage. Odysseus would have sailed through the same area but he had been warned of the rolling rocks, or drifters, by the sorceress Kirke and set another course to get around them.

"The Navahos have a similar story," Kata says. "The children of Changing Woman, the Hero Twins, come to rolling rocks on their trip to visit their father the Sun. Spider Woman gave the Twins feathers plucked from the sun bird to protect them as they passed through."

We have to debate, as one always does with Barkley and Kata, the merits of the various strategies for getting through these gates of rolling cars and trucks. We watch for breaks in the traffic, but although there are slower moments, no safe passages evolve—this is, after all, a six lane highway.

"We could simply command the traffic to halt," Barkley suggests.

"The Timucuan Indians of northern Florida used to whistle at rocks and sandbars to ask for safe passage," Kata adds. "Maybe we should try that."

The hero's task is to get around or through obstacles by any means possible. And so we hike north, nearly a half a mile out of our way, to the Boston Post Road, where we turn right, and through ingenious wile (a traditional characteristic of the hero) pass *underneath* the dangerous obstacle. To my knowledge, in all of mythology no one ever thought of such a cunning trick.

For a few hundred yards or so we hike along the road and then turn south through some scrubby woods just north of an industrial park named, with unfathomable orthography, "Lyberty Park." (Seventeenth-century Concordians were not great spellers, but they were never *that* bad.)

Merchants in these parts are especially keen to have us know that this region is the so-called cradle of democracy. Undemocratic developers, businessmen, sharp-eyed land sharks, pizza sellers, dry cleaners, and gift shop owners will inform you at every turn, through signage or company names, that we are in the land of liberty.

"Whose 'lie-berty' are they celebrating, I wonder?" Kata asks.

Never mind that eighteenth-century colonials lived under a different *Zeitgeist*—Kata and Barkley do not accept the premise of freedom and justice for all, when women, blacks, Native Americans, the landless, and debtors were excluded. Kata, in particular, has argued that the Native Americans might have been better off if the British had won the Revolutionary War. One of the causes of the war, she believes, was the fact that the British did not want the colonials to settle in territory west of the so-called Quebec Line, which belonged to the Indians (albeit not for the sake of the Indians). The colonials wanted to move westward, and furthermore they wanted the British Army to protect them against the legitimate (or so Kata says) raids of the affronted natives of the place.

Be that as it may, we are now walking through ground that was very much a part of what is known locally as the Concord Fight, and the locals are not going to let us forget it. On this road, on the morning of April 19, 1775, 131 members of the Westford Minutemen, under the command of Colonel John Robinson, marched to Concord to "defend," in the popular phrase, the town. Whether Concord needed defending is debatable.

John Robinson, who was forty years old at the time of the fight, was born in Topsfield in 1735 and lived on the western slopes of Prospect Hill. He was born to patriotism, having married one Hulda Perley of Boxford, who was the niece of the seasoned old fighter Israel Putnam, who had been scalped during the French and Indian Wars and survived to command the forces at Bunker Hill.

Sometime around two in the morning on April 19, 1775, Doctor Samuel Prescott arrived in Concord with the news that the British regulars were coming. By three or four, a messenger showed up at Robinson's house in Westford and informed him that the fateful day had arrived and the regulars were marching on Concord to search for arms. According to the recollections of Robinson's daughter, taken down when she herself was in her dotage, people began showing up at the house as he was making ready to leave. Robinson's only farewell to his wife was instructions to call up the girls and servants and have them cook provisions for the boys at the bridge. After he left, the women set about baking donuts, and they subsequently had a whole bushel sent down to Concord later that morning with a fourteen-year-old boy.

Before dawn, Robinson and his men gathered at the village green, at the muster field in front of the church. There was a brief moment of prayer for divine guidance, the so-called "artillery prayer," delivered by a patriotic clergyman named Joseph Thatcher, who was so taken with his invocation that he too marched off to join the fray. This was to be a fine little squabble, and if they meant to have a war, as the minutemen gathered at Lexington Green would be saying in just a few hours, then let it begin here.

Robinson, the ranking officer, set out for Concord on horseback. The others, 131 men of three companies, prepared to march.

By this time it was perhaps five or six in the morning, maybe later. But no doubt those living around Westford common, that is to say the Clevelands, the family of Aquila Underwood, Josiah Heald's family, and a few others, would have collected to watch the minutemen march off. It must have been enough to chill the blood—the dark figures milling and then forming into ranks, then the rattle of the drumroll, and the throbbing beat, and the fifes shrilling, and the shuffle of feet in the half-light of dawn. There is no record of this departure, but one imagines that a few of the young children and the wives followed after for some distance—down the Boston Road, past the house of John Cleveland, past Providence Meadow on the north, past the house of Eben Spalding, and down through the orchard hills, the same ones that we ourselves so recently descended—to the Four Corners of the Boston and Littleton roads, where the house of Joseph Hildreth was located.

This mustering was a scene that, to varying degrees, was being repeated throughout Middlesex County and beyond. The alarm had spread. From Acton, from Sudbury, from Lincoln and Littleton, and Groton and Carlisle and Bedford, from every Middlesex village and farm, as Longfellow would tell it one hundred years later, one by one the little companies and contingents formed up, the so-called minutemen, each with his musket and shot, ready at a minute's notice. They had been practicing for this moment for over ten years, and now it was upon them.

For our part we push on, down Lyberty Way and across the grass between the parking lots of the new industrial buildings and on into the wild forest on the southwest side of the complex. Here, in the milkweed and asters and goldenrod that the mower failed to catch, monarch butterflies are feeding.

Monarchs, along with birds, are another one of those migratory animals that would seem to prove the theory of the biological origins of devotion to place. They weigh less than an ounce each, are subject to wind, rain, cold, predators, and starvation, and yet, each summer, instead of simply giving up and dying or, like their cousins, the mourning cloaks, hibernating, they set out for Michoacán in northern Mexico. Through hurricane season, through the storms and unsettled weather of the autumnal equinox, across the whole of New England, holing up periodically at their traditional migratory stopping places, at Delaware Bay, and again at the coasts of the Gulf of Mexico itself, they fly onward, until finally, their wings ragged, their lives half spent, they reach their destination—a singular mountain slope no more than fifty acres in total. Here, they will spend the winter, and then with the warming trends of late winter and spring, they will leave for the north. The individuals who make the southward journey do not live long enough to make it back to the sunny meadows where they were born. But the genes of their progeny remember and carry on the pilgrimage in place of their parents. Other than the eels, which undertake an odyssey in which they migrate some twelve hundred miles from Europe and America to the Sargasso Sea to breed, no creature of so supposedly limited an intelligence undertakes so vast a journey to so small a place. Only a fanatical devotion to a genetically ingrained domain could generate such a marathon trip.

The evidence of this deeply rooted natural conflict between devotion to a singular place and single-minded, determined journeying is all around us today. We enter the woods and move through an area of dense shrubbery of American filbert and winterberry, black alder, and spicebush, where now ragged violet leaves cling to grassy hummocks and wild grapevines hang from branches above us. We find boletes and puffballs, and a handsome purple mushroom that none of us can identify. Once again Barkley has been halted by bird life. Hermit and gray-cheeked thrushes pass by. He spots a vesper sparrow, a swallow calls overhead, tree sparrows are moving through, as well as white-crowned sparrows and phoebes. All of them are on the move to winter quarters, having deserted the places of their birth. They will be back next spring, though. They have all of North America, more or less, to choose from, and yet they will return to a spot no larger than one or two acres. And once settled in that place, they will defend their property to the edge of death. In the nineteenth century, ornithologists believed birds defended their territories out of love. Now we have come to believe that they fight for place, for *Lebensraum*.

Mile Three

Rodrigo de Triana, lookout on the *Pinta*'s forecastle...first saw something that looked like a gleam of a white, sandy cliff standing out in the moonlight, westward, far ahead. He looked again. Now there were two white cliffs and a low dark shadow that had to be land, which seemed to connect them.

"Tierra! Tierra!" bawled Rodrigo and, with that exultant shout, disappears from history forever.

—John Bakeless, *The Eyes of Discovery*

Vine Brook 

SIX MONTHS BEFORE WE SET OUT ON OUR PILGRIMAGE to Concord, Kata and Barkley and I had undertaken our extended six-week road trip through the American South in search of the Fountain of Youth. The two of them had been seized by passionate cases of wanderlust in recent years and had begun traveling separately. Kata, in particular, seemed to suffer from an inability to stay in one place more than a few months, and in the course of a year would visit a Hopi family in Second Mesa, then travel to her parents in California, then to Peru, then back to the Berkshires for the summer months, then to Utah to camp with friends, then to Greenwich, Connecticut, to teach, and then back to Second Mesa to recover from teaching. Our journey to Florida was the first time we had traveled together in a number of years.

We began in Connecticut, swung down through the Eastern Shore of Maryland, where I still had family, and where Barkley insisted on visiting the great Blackwater Wildlife Refuge. Then, stopping often at wildlife refuges, we took obscure coastal roads along the beaches of the Eastern Seaboard until we bottomed out in Key West and the Everglades National Park, at the tip of the Florida peninsula. From there we drove north, still poking along looking for birds, to a place called Wakulla Spring in northwestern Florida, which, legend holds, was the magical fountain.

The journey was Kata's idea. She had traveled this region some years earlier on her basket research expeditions after attending a jamboree of her companion "primitive technologists," as they are called: flint knappers, basket makers, buck skinners, fashioners of bone tools, of natural brooms, combs, stone knives, and natural clay pots, who periodically collect in various locations to share their arts. Now she was anxious to go back to survey some of the areas she had failed to explore.

As we drove south, day after day Kata became increasingly disturbed that she had forgotten many of the names of the tribal people who used to hold sway in the American South, and for nearly two weeks she plagued us with the remnants of a rhyme she learned in third grade to remember the southern tribes—"Choctaw, Chickasaw, Cherokee, Creek...?" She'd forgotten the rest and, having no references at hand, became obsessed with remembering, repeating the ditty ad nauseum, to Barkley's discontent.

Kata does not necessarily recognize the "united" states of the Americas; she sees regions as tribal territories, and now that we are back in the North, near the Merrimack River in the territory of the Pawtuckets, she is attempting to construct a similar rhyme to remember the tribes of New England. Again she is without reference and again she is beginning her incessant chanting.

"Nipmuck, Pawkatuck, Massachusetts, Wassachusetts, Narrangansett...what...Wampanoag? Pequot? Doesn't work.

"Quabog, Quinnebaug, Wampanoag... Doesn't work.

"Scaticook, Pennacook, Osipee, Cree...?"

"Give over, Kata," Barkley says.

In fact it is not clear whose territory we are in at the

moment. Somewhere around here there may have been a dividing line between the people who called themselves the Pawtuckets and those who were named the Massachusetts. They might have been influenced by the powerful Nipmucks, who lived to the west; they would have had communication with the Narragansetts and the Wampanoags to the south, and also with the lesser tribal units, such as the Wassachusetts, or the Quinnebaugs. And they all lived in fear of the terrible man-eating Maquas, or Mohawks, of the region just west of the Hudson River.

Better recorded is the ending of these tribal divisions.

In the autumn of 1635 a group of Englishmen—Peter Bulkeley, Simon Willard, William Buttrick, Richard Rice, and a few others—gathered beneath an oak tree by a tributary of the river the Indians called Musketaquid—our Concord—with a powerful local leader known as Squaw Sachem, a sagamore named Tahattawan, an ally named Waban, a man the English named Nimrod, and other members of the local Massachusetts tribe. At this meeting the English purchased, for a quantity of hatchets, knives, hoes, hats, shoes, stockings, and a greatcoat, six square miles of land, a tract that Simon Willard indicated by simply pointing to the four quarters of the compass. The purchase was mere form; the land had already been granted to the petitioners by the Massachusetts legislature, the General Court, the year before.

It is not clear that Squaw Sachem and the others knew what they were getting into. Property rights, that is, outright ownership of ground, was alien to them. But no doubt they understood the concept of territory. Sections of the world, they knew, were under the control or influence of powerful figures such as Squaw

Sachem and Tahattawan, and when Willard pointed to the four quarters and announced that the English now "owned" three miles from this spot, east, west, north, and south, they got the idea. How far a mile was exactly may have been ill understood, too, but the Indians retained the hunting rights, and that was all they cared about.

When the deal was cut, a legal settlement of Englishmen was established, the place to be known as Concord. It was the first European settlement to be carved out of the inland wilderness. Why exactly the name was chosen is not clear, although it has been suggested that the founders, Peter Bulkeley and Simon Willard, wished to have an end to the nasty religious bickering that was going on among those who had fled to the New World. The name may also have been a reference to the hope for peace between Indians and English.

The origin of the name is classical. For the Romans, the word *concordia* was synonymous with the sense of community and was related to the Spanish concept of *querencia,* or place. It was also associated with civic *harmonia,* with the smooth interworking of nature and the cosmos; the personified image of Concordia was often struck on ancient coins to celebrate peace or civic agreements, and there was a temple of Concordia near the Forum in Rome. Bulkeley, Willard, and company had classical educations, so the name was no accident, but it is certainly one of the ironies of American history that a town named for peace not only should be characterized in the main by war, but should also be the place where thanks to Concord's native son, Henry Thoreau, the strategy of passive resistance was created. That dichotomy, however, has been one of the defining characteristics of Concord.

From the beginning Concord was an imaginary setting, a symbolic center wherein the English could fix their dreams. The first account of the settlement, recorded by Edward Johnson in 1636, points out that these men were the mere servants of Christ, operating according to His divine plan. In particular, Johnson says, this Peter Bulkeley, a holy man of God, went forth into the wilderness to build an inland town, which they called Concord and which was seated upon a "fair fresh river whose rivulets are filled with fresh marsh and her streams with fish." Shortly thereafter, says Johnson, breaking into verse to emphasize his points, this Peter Bulkeley, the servant of Christ, became Peter Bulkeley the warrior for Christ.

They did not come there by ordinary means, these servants of our Lord; they forged through fetid swamps, climbed tangles of downed trees, were torn and scraped by brambles and thorns, and were forced to cross scorching "desarts." Why, one wonders, did they not come by ordinary means, that is, by the well-worn Indian trails that led to and from the coast. Johnson's account does not match any realistic description of the seventeenth-century landscape reconstructed from ecological histories. For one thing, there are no sun-baked deserts this side of the Mississippi. It was clearly a symbolic journey based not on reality but on European mythic and folkloric literature. In particular it sounds a lot like the wilderness or wasteland approaches to the castle of the Fisher King, where the Holy Grail was held.

Knights errant of the Arthurian story cycles in quest of the Holy Grail had to pass through the wasteland before they could get to the castle of the ailing Fisher King. Once there, in order to attain the Grail, they had to ask a series of questions, in the proper order. If they prevailed, according to the legends, the king would be healed and the land would be restored to fruitfulness. The folkloric roots of this story predate King Arthur by a thousand years, and yet the metaphor of a sick land and a sick king and a magic cup or chalice that restores all endures even into our time. Ever since T. S. Eliot's "The Waste Land" it has become a favorite metaphor for the ailing interior landscape of the twentieth-century human psyche.

Descriptions of the New World based on imaginary literature were not uncommon in the sixteenth and seventeenth centuries; the *Zeno Narrative,* which contains the story of Sir Henry Sinclair, is a case in point. In fact, to Europeans the whole continent served as a place to center their hopes. As modern environmental interpreters have pointed out, reporters from Columbus onward wrote what they thought their readers would like to hear, not what they actually encountered.

We three in fact are probably experiencing a harder coming to Concord than Simon Willard did in the 1630s. We have been stalled by the marshes and deep waters of Vine Brook. Normally our course would take us straight southwest, but Vine Brook is in the way and once again we have to backtrack for a while, looking for a crossing. In time we come to a few high hummocks in the marshes and an old fallen tree that has formed a convenient bridge. Carefully we manage to crawl over, and then turn southwest again.

At one point in the Vine Brook marshes, we hear from an unidentifiable quarter a tiny ascending whistle, the call of the frog known as *Hyla crucifer,* the spring peeper. Normally these frogs call en masse from shallow ponds in the early spring, but a herpetologist friend of mine claims he has heard individual frogs singing in virtually every month of the year in New England, even January. The plaintive little song we hear emanating from the surrounding shrubbery is another reminder of the omnipresence of the quest theme in the literature of the Western world.

Jason assembled a great band of heroes for his voyage to Colchis to capture the Golden Fleece. One of the most powerful among them was none other than Hercules, who was traveling with a beloved armor bearer, a beautiful boy named Hylas. Shortly after the heroes left the island of Lemnos—which was inhabited entirely by women—they came to a clear spring and stopped for water. As Hylas was dipping his pitcher in the spring, a lovely water nymph spotted him and rose from the depths, attracted by his beauty. She threw her arms around him and drew him back down to her watery kingdom. Hercules became mad with the loss, forgot the quest for the fleece, and wandered off into the forest seeking the boy; the Argonauts had to sail off without him. To this day we can still hear the sweet voice of Hylas calling from the springs—even here in North America.

Kata is still working on a mnemonic chant to help her remember the tribes of the Northeast. "Micmac, Maliseet, Abenaki, Cree, Pennacook, Pequot, Wampanoag. . . ?" Barkley is spouting lines from Eliot's "The Waste Land." And I am thinking of food.

I was so busy with our various preparations that I skipped breakfast this morning and have been looking forward to an early lunch. Barkley and company are famous gourmets. They have packed a picnic of eggs mimosa, cheese, chicken *arachide* (an African specialty of Kata's involving crushed peanuts), bread, fruits, and chocolate, and by way of appetizer, Barkley has brought along some of his smoked sulphur shelf mushrooms, which he prepared at home before coming to Concord. Furthermore, since this is a celebratory pilgrimage for us—our first good walk since our Fountain of Youth venture—we are packing a bottle of Sancerre and three wineglasses. We also have, courtesy of Barkley, a small backpacker's espresso maker and an alpine camp stove. The coffee maker is one of Barkley's newest toys, and he's anxious to try it out.

Beyond the Vine Brook marshes, the land rises slightly and we come to a narrow, open field, all green with recently cut hay. South of this field there is another road that cuts between Burns Hills and the place called Parker Village. Skirting the woods, we walk along the field edge, duck back into the woods for a bit, and then, like wild animals, run across the road and jump back into the woods on the other side. A few hundred yards east of here there is an old horse trail I know about, and since we are not making good time so far, we bushwhack through brambles and pines and then turn south again on the horse trail for a little easy walking, for once. Within five minutes we are set upon by a cavalry of mounted natives.

The first sign of trouble is a routed herd of deer. Through the cleared understory of the pines, we can see, bounding abreast of us, several high-leaping does and young bucks who have been frightened by something. A minute or two later we catch sight of moving color, and hear laughter and the thud of hooves. The merry band appears and disappears between the trees, then breaks out onto the trail in front of us. Once clear of the woods, they leap into a canter and head for us, some riding in proper style, at one with their mounts, others reining in tightly, bouncing off kilter while their mad steeds prance and snort. On they come, like a troupe of wild bandits. Were we simple pilgrims in a different age, we must fall upon our knees, make the sign of the cross and lift a prayer for our safety. Innocent pilgrims on their way to Rome or Santiago de Compostela or Jerusalem were often victims of roving bands of highwaymen. We are more fortunate. The leader of this modern-day troupe raises his right hand and the group slows to a walk. We step aside to let them pass. I believe I know this crew.

"Are you from Bobby's Ranch?" I ask the leader as he jaunts by, the leather of his saddle creaking. All around now we can smell the strong odor of horse.

"We are," the leader says, and passes by, formally. He has good posture and fine leather boots, but he is wearing jeans and one of those brightly colored outdoor catalogue jackets that Barkley favors.

Bobby's Ranch is one of the two or three stables in this area that have sprung up along the Great Road, to the west of here. It caters to computer programmers and engineers who take their weekend airings on horseback, and the stables make use of any available open space north of Concord, sometimes to the displeasure of local conservationists, who accuse the horsemen of tearing up the walking trails.

The rest of the band proves more friendly.

Barkley calls out to the prettiest of the five or six horsewomen in the band. "Did you see the deer?"

"I did," she says as she jounces by with a smile. "Poor things."

"Seen those deer?" a man with narrow black eyes calls out to Barkley as he passes.

The flushed herd is a major event for this group, others mention it as they trot by.

Some of the riders are nervous; you can see fear in their eyes as they cling tightly to the saddle horn. Some are seasoned, and a few of the wiser ones seem slightly embarrassed to be out in such company. They are all dressed in jeans and bright colors and decked out in cowboy hats or tweedy peaked caps.

The noise and the smell pass, and the riders move off through the woods. We watch as they disappear and the sounds of the forest return, the popping call of a chipmunk, a woodpecker tapping, and somewhere back in the brush an airy *check* call.

"Myrtle warblers," Barkley says with finality, eyeing the bushes. (Barkley favors the older, out-of-date names for common birds: *Baltimore* orioles, as opposed to Northern, *myrtle* warblers, instead of yellow-rumped, and the like.)

This forest, according to theories of forest history, may not be so different from the forest that surrounded the inland community of Concord in 1635. The belief is that fire, insects, wind storms, and other cataclysmic events periodically destroyed large sections of woods so that young second growth was the standard, rather than the large older trees that characterize the American northwest. By the Contact Period, when Europeans and Native Americans first encountered each other, the indigenous people of New England had been practicing agriculture for as many as a thousand years, and were sustaining themselves by a combination of hunting and gathering of wild crops, combined with cultivation. According to some historians, in order to improve both the berry harvest and the deer hunting the Indians would periodically burn over sections of the land, although where, and how much exactly, is not known. Nevertheless, as we move through the woods, among the maples and the oaks of this section, we can imagine ourselves in our desired seventeenth-century landscape—provided, that is, we don't bother to check the identity of the surrounding wildflowers and shrubs, some of which came along with the alien invaders from England and Europe during the Contact Period.

The structure of the land—the hills, the courses of the streams and the marshes—is for the most part the same as well. Just to the southwest of us, we can see through the branches the outline of Nashoba Hill. This is now a fairly well known local ski slope, complete with restaurants and competitive skiing events. But before it was developed it was a place of legend. Some years back one old-timer told me with a nod of authority, as if he had been there, that the Indians used to sacrifice virgins on a large flat boulder on that hill. I later ran across this same reference in a history of the town of Littleton. In fact, the stone probably did mark some significant spiritual site for the local tribes, but what was no doubt a perfectly innocent, even venerable tradition was translated into human sacrifice by the invading Christians. As far as I know, the Eastern Woodland tribes did not need to propitiate their gods with human blood.

On the western slopes of the hill, in the place known as Quagana Hill, there was a farm held by a family from Concord named Shepard. There were three children in the Shepard family, the youngest of whom, Mary, in 1675 was a fair young woman of some fourteen years. According to the local histories, one February afternoon in 1676, during the hostilities of King Philip's War, Isaac Shepard and his two sons went out to thresh wheat in the barn at the base of Quagana Hill. Mary was posted at the summit to watch for Indians. As subsequent events indicate, Mary was a feisty, independent young woman, but she was not a good guard. Sometime in the afternoon, a small raiding band of Indians fighting in alliance with the great renegade leader Queen Weetamoo attacked the Shepard family; they killed the father and brothers and took Mary prisoner. She was carried down to Weetamoo's camp at Weninessit near present-day Mount Wachusett and imprisoned in one of the wickiups, guarded by the women or one of the warriors, possibly Weetamoo's consort, Netus. That same night, the story goes, she stole a horse and a blanket and escaped. She fled through the primeval wilderness, swam the horse across the Nashua River, and some days later arrived in Concord to report the atrocity.

By Mary Shepard's time, Concord had expanded from a village of some fifty individuals to a sizable population center. Families had grown, land had been split, and the town had been divided into three distinct sections. Northwest of the town, the great apostle to the Indians, John Eliot, had established Nashobah Plantation, one of the sixteen Christian Indian villages in the area. Eliot had managed to get the General Court to agree to grant land outright to those poor "blind" (as Eliot called them) Indians, who agreed to convert. "Conversion," by the way, did not necessarily represent a major spiritual turning for the local people. The Native Americans knew a powerful spirit being when they saw one, and this "God," the English always talked about and his son "Christ" obviously possessed some powerful medicine. On the theory that it is better to cover your bets, many accepted Christianity, which doesn't mean they rejected Hobamacho, or the spirit power Manitou, or any of the other gods in their pantheon.

When the hostilities of King Philip's War finally broke out, as was probably inevitable, the proximity of Indians, Christian or otherwise, was problematic for the local Concordians. During the war itself, when the men were off fighting the Indians, the women and children would commonly set up watches and race to the nearest fortified house when an alarm went up. These were not technically garrison houses—when Concord was founded, the locals had no fear of the local Indians—and the nearest fort was constructed on the eastern slopes of Nashoba Hill in the latter half of the century, not far from the ill-fated Shepard family farm. The house had a good well, which is still there, and a small tributary of Vine Brook ran in front of the place, also still visible.

In 1676, the problem of possibly dangerous local Indians at Nashoba was resolved. The English rounded up the Indians of Nashobah Plantation and marched them down to Concord, where they were ensconced in a workshop and stockade near the place that was later to be

Bronson Alcott's residence, Orchard House. This proximity made the residents even more nervous, and one day, on no one's authority, a feisty Captain John Moseley from Marlborough, who had fought at the nasty battle at Deerfield earlier that year, collected the Indian men, women, and children and had his soldiers escort them to Deer Island in Boston Harbor, where other Indians, including converted Christians, were imprisoned. It was February. It was cold. There wasn't much to eat, and of the fifty or so peaceable souls, only two or three survived to come back to the Nashoba region after the hostilities ended.

One Indian somehow managed to avoid imprisonment, however. The man the English called Tom Dublet kept a fish weir on Beaver Brook, which winds below Prospect Hill and eventually empties into Stony Brook and the Merrimack River. This was the stream that Sinclair reportedly ascended when he was exploring the interior of New England in 1398. Dublet's father had a weir on this stream and in fact was killed there in 1630 during a raid by the Maqua Indians. His son Tom was Christianized by John Eliot and was probably a member of the Nashobah Plantation. Tom married the granddaughter of the sagamore, Tahattawan, one of the leaders who sold Concord to the whites.

Early in the war, the Indians learned that these English placed a great value on women and children and would go so far as to trade food and guns in order to have them returned. So the natives began taking captives—normally they would have killed most of their enemies, no matter what their age or sex.

On February 10, 1676, there was a well-documented raid on the outlying town of Lancaster. Indians attacked the town at dawn and began setting fire to the houses. A few families managed to take shelter in the house of the minister of the town, the Reverend Joseph Rowlandson, but the Indians surrounded the place and set it ablaze as well. As the flames rose, the painted native warriors circled and jeered at the plight of the cowering English, and whenever people would emerge to put out the flames, they were showered with arrows. Inside, the people were terrorized; even the six guard dogs, who had been trained to despise Indians, were uncharacteristically cowed by the raid. At one point Rowlandson's wife, Mary, came to the door and was wounded by an arrow, as was her six-year-old-daughter. The flames were licking the sides of the house by this time, and finally, rather than roast, the English came out. Most of them were either killed or captured. What's more, all the cattle were slaughtered, a fact that seemed to bother Mary Rowlandson more than the death of some of her compatriots: the Puritans did not appreciate waste.

Only one person escaped from the attack. He fled to homeport, Concord, to report the crime. The Concordians raised a company of fifteen foot soldiers and marched to Lancaster, where they found a few survivors and many scattered, horribly mutilated bodies. Mary and her children were not there. They had been taken captive for ransom.

Mary kept good notes of her captivity and redemption. The journal is a desperate account of hardship, suffered not only by the captives, but also by the captors—the Indians themselves had been pushed to the brink of starvation by the privations of this war. Their idea had been to push the English back into the sea and send them on to England in their ships. But the disruption of the normal hunting and planting cycles caused by the state of war was wreaking havoc among them. The Indians were familiar with raids, not prolonged wars.

After two months on the move, avoiding companies of English and searching for food, the Indians sent word that they would negotiate an exchange. The English needed someone who understood both sides in the affair and selected Tom Dublet to undertake the proceedings. He traveled out to the Wachusetts area twice before he managed to gain the release of Mary Rowlandson and her children.

When the war was over Dublet went back to his fishing weir on Beaver Brook and lived out his days with a woman named Sarah, who was either his wife or his sister or daughter and who lived on to become the last surviving Indian in the Concord region. Local histories describe Dublet as a tractable old Indian. But in fact he must have been embittered, perhaps a bit crazed, by his experiences. The English had no respect for him. Most of his Christianized Indian allies were dead of starvation, and the rebellious natives must have despised him for his complicity. He spent the last fifteen years of his life suing the General Court for some recompense for his negotiating efforts. In the end the Court granted him an award: a red waistcoat with shiny brass buttons.

The insult may have driven him mad. There is a document in which one Tom Dublet is accused of dallying with a white girl on the marshes of Beaver Brook. Though the old man was taken to court, he was exonerated—the court claiming that the young girl was a willing accomplice—but there are rumors, as there often are in local histories of this sort, of a curse. Dublet may have had shamanistic powers, it turns out.

If it is true, as Kata says, that mythic events are played out in everyday life, then it is possible that Dublet's curse is still at work in our time, disguised by the changing economies of computer technologies.

The great renting highway, Route 495, which so successfully blocked our natural passage, runs just east of the bend in Beaver Brook where Dublet had his fish weir and where he lived out his last days in a small circular hut. The highway intersects at that point with the Great Road, a former Indian trail that runs down to Concord from the hinterlands and hill country of southern New Hampshire. The intersection is a fine spot for industry, located as it is not two miles from Route 2, an east-west state highway, and, about ten miles to the north, Routes 3, 4, and 93.

There was an excellent farm just east of Dublet's weir that was planted with some fifty species of trees but was subsequently destroyed by the highway. For a while one of the local stables used to graze horses on the land, and wandering tourists, taking the scenic route up the Great Road, would stop to feed and photograph them. Then, in the 1970s, land speculators sold the place to the GenRad Corporation, which planned to construct a mighty industrial fortress in the area. Before construction could begin, however, there was a curious downturn in the economy, the grand scheme fell through, and the land went back on the market.

It languished for years. Foxes moved in. I once found the fresh body of a screech owl there, a bird sacred to the people of Tom Dublet. The lanes of the old farm, the trees, the bird life, attracted people to the place in spite of the proximity of the highway. And then, we innocent locals—who exist in our chosen place at the mercy of land-hungry developers, hoping against hope that the land will remain as it is and life will go on—learned that yet another massive industrial complex was about to descend.

A few rebellious natives rose up and went to interminable meetings to try to stop or control the development. The associate editor of the local paper did all he could, without revealing his true opinion, to expose the folly of the development. He even went so far as to print on April first a mock edition of the paper, in which he reported that the developer had met with a local shaman and had been converted to conservation by the magical intervention of a spirit bear and was intending to repent his sinful career as a developer.

To his credit, the developer enjoyed the joke. But his bankers were not amused. When the developer appeared at a loan office meeting shortly after the paper came out, he was greeted by a circle of stony-faced accountants. When he finally was able to summon the courage to ask what had gone wrong, one of the creditors threw the paper onto the table without a word. It took the developer the rest of the meeting to explain the concept of April Fools' Day.

Things went downhill after that, albeit for different reasons. The developer considered himself something of a visionary. He hired the best landscape architectural and planning firm in the East to create a site that would be attractive to expensive, upscale computer companies. The planners envisioned fountains and walkways open to the public, groves of exotic trees, sensitively placed edifices, ponds, preserved wetlands, traditionally lighted drives.

Day after day bulldozers worked the land, evicting foxes and owls, blue jays, and the cedar waxwings that used to pass over the area. The drives were laid out, well-crafted stone walls were constructed, lamps were installed, and finally, after nearly a year of site work, a sign went up announcing the arrival of "Littleton Corporate Common." Never mind that the traditional New England common, a child of the European village and cathedral plaza, was the very anchor of the sense of place of towns and cities. Never mind that this development would effectively obliterate any remnant of the sense of the place. The plans for the grand corporate center went on.

By this time the developer had read some of the literature concerning Tom Dublet and his fish weir and the presence of a Pawtucket spirit bear who haunted the land to the west of the weir and the proposed development. In jest, he asked that some of the New Age shamans who had been attracted to the site by its mythic overtones intercede on his behalf and keep the bear spirit and Tom Dublet at bay. But it was too late. Suddenly the Massachusetts miracle began to come apart: office spaces began to empty, cutting-edge computer technology flourished elsewhere, and the dream was over.

Now weeds grow along the landscaped drives. Some of the lamps have broken, blue jays have returned, and one night I saw a fox dash across the Great Road and head up the driveway, between the finely built stone walls. It was either cosmic coincidence, or Dublet had struck again.

South of Nashoba Hill, Vine Brook, which flows southeast along the general route taken by the Westford Minutemen, joins Nashoba Brook, which runs by the Walter Powers house, used as a fort during King Philip's War. At the point where the two brooks join there is a wide, impenetrable marsh of reed canary grass interspersed with speckled alder thickets and many deep little tributaries. It was streams and marshes of this sort that caused the 131 men of Colonel John Robinson's company to take a meandering course on their way to the Concord Bridge fight. Bridges were few in those days, the land was wild, and the streams were many.

In our time, if you attempt to walk directly to the North Bridge from Westford, you will encounter the same problem. We are at present knee-deep in thicket and casting about the marsh looking for a crossing. The ground has broken up into a mazework of grassy, reedy hummocks, with black ooze or standing water between them. We leap from hillock to hillock, trying to get closer to the main stream, but are periodically forced back, and when we finally get to Nashoba Brook, it is too wide for us to jump and too deep to wade, even if we were so inclined, and is uncrossed by convenient logs, as was Vine Brook.

Barkley is studying our geological survey map intensely, looking for some sort of unbuilt high ground where we can get over. While we are studying our possible route, Kata comes over, looks at the map briefly, jabs a finger at a point of land and says, "Let's go there."

It is a new housing development just to the north, on high ground, with a bridge over the brook.

MILE FOUR

Ladies sad will lose their mates,
The land in desolation lie,
Damsels unconsoled will sigh,
Widows and orphans, mournful all,
And many a knight in death will fall.

—Chrétien de Troyes, *Perceval*

Nonset Brook 

We walk along the brook for a while, still looking for a crossing, and then, finding no way over, we backtrack to high ground and climb up the banks into a woods of white pine and highbush blueberry. All this was pasture once; we can see the well-built stone walls winding through the trees and we follow one to a clearing. Just beyond the trees, across a small lawn, is a comfortable suburban house with a neat stack of firewood beside the kitchen door. We see a child's swimming pool near the woods, a few plastic toys in the yard, a rake leaning against the house, and, strung between the east wall and the nearest tree in the woods, an omen—a heavy wire dog run with a large chain hanging from it.

"Dare we?" Kata asks eyeing the dog wire.

We are crouching like raiding Pawtuckets in the shrubbery.

"He's not there," Barkley says. "The chain is empty."

"I know, but look at the size of it. What if he's around the front."

The chain is indeed large and there is an immense dog bowl by the house and freshly dug-up ground.

"Another pit bull," says Barkley.

We had had several incidents with unfriendly pit bulls and Rottweilers during our recent pilgrimage to the Fountain of Youth.

There is another house farther down the street, and just as we are about to retreat and come out of the woods there, the dreaded beast emerges from under the porch. It is an overweight golden retriever, its feathery tail waving gladly in the autumn air and a disgusting old tennis ball in its jaws. Unceremoniously it deposits the ball at Barkley's feet.

We straggle out onto one of the suburban streets, our muddy boots leaving little clumps of soil behind us. We pass well-tended lawns and clean, asphalted driveways, and, just after the bridge over Nashoba Brook, cross a side street with a road sign that informs us that we are now on White Pine Drive and are passing Lady Slipper Lane. Why is it, we wonder, that developers feel it necessary to name streets after that which they have successfully destroyed? My favorite is a street in a new development not far from here called Preservation Way.

We must seem a motley crew by now; Barkley has a scraped cheek, Kata's hair is loosened, and we stride down the middle of White Pine Drive as if we knew this place. In fact we are lost. The road dead-ends at a trail, which we follow for a while, heading eastward in the general direction of Concord. The land at this point is impoverished—gravelly soils, the trail lined with sweet ferns and goldenrod and asters—and there are signs of horses everywhere, droppings and hoof prints intertwined with the ruts of mountain bikes and dirt bikes. I believe we are coming to a quarry I know of not far from an area known as the Nashoba Brook Wildlife Sanctuary, owned by the Massachusetts Audubon Society, but the approach is unfamiliar. Suddenly, ahead of us, we hear gunfire, so we halt again to reconnoiter.

After some discussion we decide to move forward and investigate, and as I expected, in time we come to an open area with a quarry pond. In the distance, near a gravel bank, we can see three figures. Barkley eyes them with his binoculars and determines that they are inspecting something at a gravel bank. Taking our hearts in our hands, we decide to approach to ask them directions. Barkley and Kata love to ask directions, sometimes even when they know where they are.

Our resident knight-errant, Sir Barkley, has had the Grail quest in mind all day as we slouched eastward towards Concord. But we are all having trouble deciding which of the many habitats that we have passed through so far represents the wasteland. I hold that it is the shopping malls at Four Corners in Westford, the hideous, mindless, flat, brightly lit, unvegetated, dry, desertlike environment of commercial blight that has ruined the American village, obliterated any sense of place, and caused the destruction of community in this country. Kata believes it is the strung-out, lonely, private dwellings we have just passed, the anonymous, isolated suburbs in which television (a recognized wasteland in itself) has replaced community and has become the core of the sacred *domus,* or house, an electronic version of the house altars of ancient Rome, of Buddhist domestic shrines, Confucian family altars, the *retablos* of Mexican peasant households, or the decorated household pujas of Hindu dwellings.

Barkley claims that the land we are now passing through is in fact a real wasteland, not a metaphorical one. Gravel trucks have attacked this place. For two hundred acres or more the land in front of us has been stripped of its original vegetation: the oaks, maples, elms, white ash;

the laurel, blueberries, viburnum, the grapevines; the filbert, royal fern, bracken, ground cedar; the polypody ferns, mosses, the very lichens of the rocks—all have been ruthlessly torn from the native ground. Machines have dug at the very foundation of the earth, raked away the ten thousand-year-old topsoil and dug down into the ancient layers of gravel deposited by the glacier some fifteen thousand years ago. Now only a few pioneer plants have returned—poverty grass, sweetfern, a few asters. It is a tragic contrast to the marsh we have just passed through and the wooded hillside that we can see beyond this pit of hell. Furthermore, like all the approaches to the Grail Castle, this is a dangerous country to be in. As we draw near to the distant figures, we can see that they are armed.

The three people crouched beside the embankment are staring at something white on the ground. One of them, an armed Amazon, stands a few paces back from the hunched males, and when they stand up and move apart, Barkley can see through his binoculars that they are inspecting a target.

"Not dangerous," he pronounces. "They are not in their killing mode."

They turn out to be friendly natives. The two men are mustachioed, smoke Marlboros, and are dressed in jeans and shiny jackets that advertise their names, Frank and Troy. Their companion, Mary, has frosted blond hair and wears a loosely knit sweater. She and Troy, we later guess, must be married. All of them are well armed. Troy and Frank have leather belts and holsters with large, exposed pistol handles, and Mary has a small twenty-two-caliber rifle crooked in her arm.

We explain our mission, show them the map, and ask

for the best route to get over to the Nashoba Brook Wildlife Sanctuary, which is just to the northeast. Mary takes over immediately and enthusiastically. Troy is not so certain. He lights a cigarette while she describes a route, and I spot, I believe, a glance of exasperation, a certain condescending tolerance.

"I don't think that's the way exactly, Mary," he says when she is finished with her explanation. "They would do better to just cut over through the hill there. They'll hit a trail back in the woods."

"They go over back there, they'll be on that same trail," she says, "only easier walking."

"Right, but it's the long way around."

"They got all this puckerbrush if they go your way."

"They don't mind bushwhacking, Mary. That's what they're doing."

"It's not that far, it's no big deal, Troy. They won't get scratched up."

Troy takes a draw on his cigarette and stares off into the middle distance.

I know exactly what Barkley is thinking now. We go either way, we are going to be shot in the back because we have failed to obey one or the other.

"Well, it's a fine day, in any case," Barkley says. "Maybe we'll just take the least confusing route so we don't get lost again."

"Well, you just go back a little then," Mary says. "Turn right where you see the cut in the trees. That's the trail, take you right into the sanctuary."

"You all can't hunt in there?" I ask, knowing full well they can't but that deer hunters have been invading the sanctuary.

"You can't hunt nowhere anymore," Frank says, breaking an ominous silence. "Too many horsy types around here."

The woods of the sanctuary offer easier walking—sharp little hills, eskers, some flat lands—and finally, another fine little brook, known as Nonset, which eventually feeds into Vine and Nashoba brooks. At a pleasant embankment above Nonset we decide to take morning coffee. Barkley extracts his equipage, fills his new espresso maker with water, tamps down the coffee, and lights his small, gas-fired pack stove.

Somewhere around here we may have passed a divide. At this point, Nonset Brook flows south-southeast. Beaver Brook, the main stream of these parts, flows into Forge Pond and then becomes Stony Brook, which is a tributary of the Merrimack River; the Merrimack collects in the highlands of New Hampshire and flows down to Massachusetts Bay and the sea at Salisbury, gathering waters all the way. East of Westford, beyond Sparks Hill, Burns Hill, and Nashoba Hill, the streams either run between wetlands or, as we are doing, head by the best means possible to the Concord River, which, because of a phenomenon known as glacial rebound, flows north, also to the Merrimack. The ice of the late great glacier was so heavy it compressed the very earth, and when the glacier departed (for the time being—we are in an interglacial period), the land rose behind it, creating a general north-running slope.

This glacier is at once the source of all our problems on this pilgrimage, and the savior of the land we are journeying through. It is the reason there are so many streams and wetlands in this particular patch of earth. And because the state of Massachusetts has a law that in theory at least, protects wetlands, it is the reason so much of the land between Westford and Concord is undeveloped, allowing us more or less wild passage to our intended destination.

There is a a red maple just across the stream from us, its low branches sweeping the water. Beyond the tree we can see higher ground, a black cherry, a wide-spreading white oak, and some young chestnut trees. Just across the brook a stone wall emerges from the woods and ends at the bank, and here it has been fashioned into a stone alcove with a bench. Years ago, when the land beyond the brook was pasture, this was someone's retreat. We can imagine some winsome dairy maid dangling her bare feet in the dappled waters, and perhaps of a rare June evening the squire himself would break off from his interminable labors, come to this spot, and admire his stonework, the wooded bank across the stream, and the last songs of the wood thrush.

To the early Greeks, springs and wooded stream banks of this sort were sacred. Nymphs and naiads lingered in such spots, gods came down to earth in these places for terrestrial trysts, demigods and heroes were conceived. As late as the nineteen-fifties, Greek peasants would leave offerings at such sacred springs and groves—oranges and figs and grapes.

Waters are often the very essence of the sacred places of the world. At the heart of Mecca, in the very core of the Great Mosque, there is a sacred well, the Zamzam, whose waters are fresh and cold and have miraculous powers. It is said that even clothes washed in the waters of the Zamzam have the spirit within them;

devout Moslems prefer to be buried in garments that have been washed in its waters. The spring was a pilgrimage site for desert tribes a thousand years before the Moslems took it over.

In 1858, a sickly French peasant girl named Bernadette Soubirous saw a vision of a lady in white in a cave above a stream along the sacred route to the great medieval pilgrimage site Santiago de Compostela in northern Spain. Bernadette called the lady in white "Aquero," the one who is there; it was not until later that church authorities in Lourdes pointed out to her that this was the Virgin Mary. In one of her appearances "Aquero" told Bernadette to dig at a spot beside the stream in front of the grotto. A spring welled up whose waters are believed to be curative. Now three million pilgrims descend on the spot each year.

Traditional Chinese would say these places have strong *feng shui,* a powerful, potentially beneficent force that occurs in specific areas, a sort of spirit of place or ambience that must be carefully reckoned with. *Feng shui,* literally "wind/water," is associated with the Chinese concept of *chi,* the life force that circulates throughout the earth and also within the human body and pours out of the earth in certain places around the world.

Barkley's machine begins to splutter and a thin dark stream of thick coffee streams out into a china demitasse Kata has packed. Barkley passes her the cup and she drinks from it dreamily, while he prepares a second. There is a lot of ritual involved in the operation. The top of the machine has to be unscrewed, the grounds dumped, the cup repacked, and the small pot filled with water and placed again on the stove. But coffee is the elixir of this life, says Barkley. He and Kata never travel without their own supply.

There was an early medieval myth, probably of Celtic origin or perhaps brought to northern Europe with the Romans, who may have picked it up from the Greeks, of a magic spring or fountain. The spring was hidden from the common dross, but humble pilgrims, knights, and spiritual questers of one species or another could find it. If you drank from the waters of this spring, your youth would be restored and you would gain eternal life. The Grail, the unending source of *chi* and sustenance, was a part of this tradition.

According to the Spanish historian Herrera y Tordesillas, in 1512, when Ponce de León was serving as governor of Puerto Rico, soldiers brought to him an old wizened Carib woman, *una vieja,* from the hills of Luquillo who told him of a nearby island where there was a wonderful spring called Bimini, which gushed up out of a deep fissure in the earth and had the power to restore youth to aged people who bathed in it. Ponce de León was over fifty at the time, not a bad age for a ruthless conquistador. He had arrived with the second voyage of Columbus in 1493, had been a soldier with the fiendish Nicolás de Ovando on the island of Hispaniola, and had made a name for himself as an Indian fighter before being named governor. Juan Ponce de León was out of the noble house of León. For generations his family had fought Moors, and one of his relatives had been Duke of

Cadiz. Juan himself had been a soldier since the age of thirteen and knew a thing or two about warfare. By 1513, having strewn so much death and destruction around him, he must have been feeling the pull of mortality, so he had the *vieja* confined for further questioning, and after some time decided to make a voyage through the Bahamas chain to see if he could find the island.

On March 3, leaving his wife and two daughters at the Casa Blanca, the castle he had built for himself on Puerto Rico, he outfitted an expedition of three ships and took a northeasterly course through the Mona Passage. He coasted along the northern shore of Hispaniola and then continued northwest, threading his way through some six hundred miles of the treacherous waters of the Bahamian archipelago. Somewhere near Grand Bahama Island he turned west and sailed out into the unknown.

All this has the ring of myth about it, and in fact it probably isn't true. The first reports of anything resembling a fountain of youth in the Americas were fostered by a young Spaniard named Fontaneda, who had been captured by the Indians and had heard a local legend of a river that restored youth. Ponce de León's earliest biographer, Herrera, ran across the account and associated the search for the sacred waters with the avaricious Ponce de León, who was probably searching for more gold when he happened upon the mainland of America. Herrera must have known that the legend would find a sympathetic readership in sixteenth-century Europe. The myth of the magic spring was well ensconced in the European psyche.

Fountains of youth or springs of sacred waters are traditionally associated with wise crones and new lands.

There is, for example, an Irish legend of five brothers who develop a great thirst while hunting and come upon a well of sweet water guarded by a hideous crone with fangs and scraggly hair. When she demands a kiss before granting the waters, the first four brothers are so repulsed they come away dry. But the last, Nial, not only kisses but embraces the old woman, who subsequently is transformed into a beautiful demoiselle and grants him dominion over her land.

Barkley's espresso machine spits out its second cup of coffee. He passes the cup to me and begins again. He is well satisfied, I can tell, by his new machine, and in fact the coffee is a fine complement to the autumn air. The weather is cool but clear and, in the sunny spots, actually warm.

The old crone had told Ponce de León that the island of the sacred spring lay to the northwest, so after stopping at a few of the outlying Lucayos, or Bahamas, and finding no springs, he sailed on. His pilot had taken him through a labyrinth of coral reefs, and on into the Gulf Stream, and in time the three caravels came to what they thought was a larger island. They turned north and coasted up the beach, sailing just beyond the white line of surf until they came to a great river. Somewhere in this passage, they landed. It was Easter Sunday, *Pascua Florida,* the height of the southern spring, and the forests were all abloom. He called the place La Florida.

Ponce de León's northernmost landing in Florida

was probably somewhere around the St. John's River. But having found no Fountain of Youth, or no gold, whichever you prefer, he turned his expedition around and, fighting adverse seas and winds all the way, sailed back down the coast, rounding Cape Canaveral and struggling for weeks against the north-flowing Gulf Stream. He came eventually to the Keys, sailed along them, and turned west until he reached a small group of islands where he found no spring, but thousands upon thousands of green sea turtles. He named the islands for them, Tortugas. From there he turned north again, to Cape Sable and the mangrove-tormented coast of the Everglades. One of his vessels was in need of repair, they were low on firewood, the waters were shallow and treacherous, and every time they came near to a good landing, the implacable natives appeared with their vicious three-pronged arrows. Ponce de León finally managed to careen his vessel on one of the keys, made the necessary repairs, and sailed on. For days and weeks he hugged the coast, stopping at mangrove keys where white birds filled every branch, where great diving pelicans soared in the air, and where he saw storks and egrets, and ibis, and strange birds with bills shaped like spoons. But no fountains of youth, and hardly a good harbor. He finally turned and sailed back down the peninsula and retraced his course across the Straits of Florida to the Bahamas, where he stopped at an island he later named Bimini. On a nearby island they called Vieja, they found a lone old woman and questioned her. She too knew of the sacred well. But Juan Ponce de León was weary with sailing and in need of rest, so he returned to Puerto Rico, to regroup.

Barkley rewards himself with the third cup. He is a generous host, always seeing to your well-being, bearing gifts and good wines and stories and adventures. Kata, by contrast, prepares certain special dishes and tends to make gifts for you rather than buy them. In fact, she buys hardly anything, except books, of which she has thousands. But that is not to say she does not acquire things. The house of Barkley and Kata is like a nineteenth-century museum storeroom of artifacts, most of them Kata's. She cannot resist collecting basket material wherever she travels. Some rooms of their house are devoted entirely to ash splints, wisteria vines, oak strips, black ash strips, the stacked barks of a variety of trees, knives, vine ties, splitting mauls, wedges, canoe frames, leaves of wild teas, baskets of her own making, baskets of known and unknown weavers, and of course books, books on natural history, on trees, on snakes, bears, flowers, books on Indians—the Ojibway, the Blackfoot, the Crow, the Paiute, and the Pawnee. Books she will never read, on theories of European history, books she reads again and again, poems and ritual orations, mystic writings. She has saved postcards of Indians, bears, wolves, indigenous peoples from all parts, landscapes from around the world. Nothing that relates to earth remains unrepresented.

Stacked in an upstairs room, rarely visible to visitors, is Barkley's gear, his cherished copy of *The Divine Comedy*, illustrated by Gustave Doré, current fiction, works of travel writers, a camera, packs, boots, and clothes for traveling, as if he should be ready to leave for China on half a day's notice—something he has in fact done, I believe.

On our trip to find the Fountain of Youth, Barkley had stowed his small pack neatly in the trunk of his twenty-year-old Mercedes-Benz. Kata brought along a bag of rice, a stash of dried wild mushrooms, dried peas, cumin, curry powder, her flute, some camping gear, pots, water, Mary Theilgaard Watts's *Reading the Landscape of America,* Bartram's *Travels,* and other accounts of the American South, including a book called *Tombee,* about Sea Island plantation-slave life. We crossed the Delaware River in a sleet storm, suspended high above the gray waters as if in flight, Barkley driving and Kata playing her flute and starting to recite her list of tribes of the South. That night we stayed on the Eastern Shore and I visited relatives while Barkley scouted for waterfowl at the Blackwater Refuge. The following night we stayed at a sportsman's lodge near Swann's Quarter—still raining, waterfowl everywhere. In the morning, while we were eating breakfast, we saw an otter swim by. Barkley, needless to say, was in heaven, in spite of the weather. We were determined, or at least Kata was determined, to find the apocryphal site of the Fountain of Youth. But as usual with them, we were in it for the trip, not the destination.

In some ways our walk to Concord is the last leg of this earlier pilgrimage, and the two parts are providing a significant dialectic. When we went south, in keeping with Barkley and Kata's penchant for exploration, we took the back roads all the way to the Everglades. But one does not travel anywhere in America without coming up against the reality of strip development, and time and again we ran into the latest iteration of the American dream, the late-twentieth-century version of Henry Miller's air-conditioned nightmare. Barkley manages to get through such areas without suffering; he seems to have the ability to intellectualize entire landscapes and sets himself up as a sort of Virgil, guiding us through hell. Kata, by contrast, is easily overwhelmed. The commercial edge cities at the highway interchanges, the standardized repetition of signs, architecture, food, landscape, send her into periodic tirades in which she begins with one problem—decaying urban centers, for example—and then launches into a litany of other ills: historic sites razed, arable farmlands destroyed for shopping malls, forests stripped, wildlife extirpated, species threatened or lost entirely, native cultures destroyed, mile upon mile of fast food outlets, shopping centers with glaring lights, violence in the streets, death penalties, too much TV, millions homeless, popular music consisting of nothing more than noise and rhythm, serious music gone awry with cacophony, art broken into solipsistic masturbation, people molded by a common language, and on and on.

"If the landscape around here is any indication of the direction this country is taking, we're in serious trouble," she says. "And here we are, risking our lives, driving through murderous Florida. What is the point? What are we doing here?" she asks.

"It was your idea," says Barkley quietly.

For Barkley the landscape of malls and strip development is the new vernacular landscape of America, the folkloric expression of the people's choice; if the people did not want them, he says, developers would build something else. He takes these places for what they are and judges them according to individual expressions. Some malls, he thinks, work for a particular place, some are failures. Some strip developments he likes, some he hates. At one point, somewhere in south Florida, he made us stop at an American baseball game taking place under brilliant lights, the players suspended, as he expressed it, "like fish in an aquarium." The field had an eerie green cast to it and the lights hurt Kata's eyes, but we stopped anyway.

Often, normal people do not know what Barkley is talking about; he speaks in metaphor and presumes everyone knows what he is saying. You have to know him to excuse him. A few days after the baseball game he said something about Kant's nature of reality to an Everglades park ranger, which even Kata and I did not understand. The ranger smiled and nodded as if she understood—she must have been trained to deal with kooks—and later was forgiving enough to join us at a quasi-Mexican restaurant in Homestead. Barkley's choice, needless to say.

The strip development closest to Concord is the so-called Great Road, where you will find, in a ten-mile stretch, one stop light, one Burger King, three auto sales lots, three gas stations, three small, tastefully lit shopping plazas, two French restaurants (both housed in eighteenth-century buildings), a small wooded reservoir providing Concord's water, and, beyond, just over the Littleton line, horse farms, a small housing development, and, lined up one after the other like the fast food joints of less fortunate communities, five farm stands, their wide hayfields and market gardens rolling out behind them. It is not an unpleasant drive, and it is even listed as a "scenic road" on state highway maps. But since I live near the Great Road, I consider this stretch one of the greatest eyesores in America. I have to leave town to put it in context.

One has to go to Florida to truly understand America. The state contains within its borders the totality of the North American experience. In any given winter, a large portion of that element of the U.S. and Canadian populations that is not within striking distance of southern California drains down to that green, subtropical peninsula that for the last five hundred years has drawn the dreamers of eternal youth. Ponce de León's chronicler had the right idea. It's not gold people want, it's pleasure, which is perhaps healthier than the pursuit of wealth and power. But the pursuit of this goal has ruined one of the most diverse ecosystems on the continent.

We pack up our coffee, cross the Nonset Brook, and start again. Almost immediately we again come up against impenetrable wetlands, and rather than backtrack to the sand and gravel beds, where our armed guides are still target shooting, we hike northeast for a while to try to cross Butter Brook at what looks like a narrow spot. Halfway there, in the middle of a stand of arrowwood and tupelo saplings, Barkley raises his hand like a cautious African guide.

"Listen," he says. "Finches."

Above the trees we hear a bright chirping.

"Evening grosbeaks. Very early arrivals, I would say."

Evening grosbeaks are northern birds who periodically come down from the spruce belt of northern Canada to spend the winter in New England. The flock passes overhead, moving in the direction of the sand and gravel beds.

We carry on through an upland woods, cross a small hayfield with houses on the opposite side, and enter the woods again. At one point, quite suddenly, we burst out of the brush into the backyard of someone's old farmhouse.

This is the type of place that once characterized much of rural New England. It is sheathed in unpainted clapboards, there are several broken windows covered with plastic, the cedar shingles on the roof need replacing, and there is a canted oil tank next to a backdoor, which appears to have been permanently nailed shut. The backyard consists of blackberry brambles that run up to the tank, and there is an old car in the woods just north of the house, a great gray Chevrolet, rusted and moss-covered, with the hood popped open. In some ways it is a relief to see such a ruinous place here among the estates of the wealthy horse country of Westford or Carlisle, or wherever we are at the moment.

Just as we are about to back out of the brambles and return to the forest, we see an old man with a rake and overalls standing waist-deep in blackberries. His mouth is opened slightly and, under his red-patched hunter's cap, its ear flaps tied on top with lengths of string, his eyes stare blankly. We greet him apologetically.

"On our way to Concord," Barkley says.

"You'll just be taking the Old Road then," he says. "Come on through." He motions with his rake.

This is not what we want, but we weave through the brambles and join him in a cleared area where he has his wood pile.

"Car break down?" he asks. "Or are you bird hunters?"

We try to explain our mission but I can tell he doesn't really care; he's happy for the chance to talk, and far be it from us not to oblige him.

It turns out we have nearly rejoined the march route of Colonel Robinson's minutemen. The company, having marched down the Carlisle Road through the section of Westford known as Parker Village, turned south here on a dirt drive now known as the Old Road, which in 1775 was the main route to the Concord area. The old man seems to know the story, as if he had been a young lad on that fateful morning when the company passed in front of his dooryard. His house, I would guess from the architectural details, might have been standing on that day.

He turns out to be a member of one of these families that occur in Concord and its environs who get a piece of farmland and cannot quite grasp the concept of ever leaving. He is from a large Irish family, well known in the area, whose people have worked their way into the folklore and rumor mills of the town. One of the brothers was a famous dump keeper who brought so many treasures home that he in effect turned his yard into the town dump. Another slipped into alcoholism and died one night while sleeping in the local graveyard. For years the old man's father raised crops on this property; the old man followed his father's path, but slowly lost ground to blackberries.

"Was married once, but it never did seem right to go

to Florida," he says cryptically, then slowly pays out the story. Late in life, after she suffered a heart attack, his wife got the idea that she would like to visit, at least once, the mythic state of Florida. He was not keen on the journey, for one thing he had never been much beyond Boston, but he went along with the idea, and in February they started south, taking the interstates, which in itself was a bit of an ordeal.

"There were a lot of signs to read, and you have to be able to read *fast.*"

Somewhere around New Jersey he began worrying about his beehives.

"Was afraid a wind would take down a branch, don't you know, knock the hives askew and freeze them all. February is not a good month to leave home."

They pushed on, through Virginia, North Carolina, and into the motels near Brunswick, Georgia.

"Wife wakes up in the night with pains in her chest and says, 'Let's go.' I thought she meant to the hospital, but no, she wants to go to Florida before she dies. Never could argue with her so we drove on. Four or five in the morning. About dawn we saw the signs saying 'Welcome to Florida' and she says, all right, take me to the hospital. Next day, she died."

"But this is terrible," Kata says, touching his arm. "What a terrible journey for you, and you didn't even want to go."

"Well, it was a long time ago, and I got some new bees."

"And you carried on, here?" Kata says,

"Well I used to have more land, but I had to let it go."

"You sold, you mean?"

"No, I mean I had to let the woods come back. Got too old to cultivate."

"What if you sold?" Barkley asks.

"Sell? Sell what, the land?"

"Yes. You could probably get a good price for twenty acres or so."

"Oh I couldn't do that," he says. "I couldn't sell the land. You've got to have land."

This is the type of man the local real estate agents cannot comprehend. There are others like him in the Concord area. Surrounding the town are rich farmlands, some of which have been cultivated since the 1640s. One farm in particular until very recently was in the same family since 1666. The farmlands were once divided along ethnic lines and still are to some extent, the Italians to the east, the Irish to the north, Swedes mixed in among them. Real estate values have soared in Concord in recent years. Southeast of us, on Monument Street, land sells for as much as two hundred thousand dollars an acre—without a house on it—and some of the Concord farms have as many as two or three hundred acres and are owned by poor families who get up at dawn to weed or milk cows and work till dusk, and carry on seven days a week in this manner during the nine-month growing season. I once asked one of the older Italian farmers, who was particularly hard-pressed financially, why the family didn't sell. "We wouldn't know what to do," he said. "What are we going to do if we don't grow vegetables? Become hairdressers?"

Mile Five

Upon their Emergence to the new Fourth World, the people were told that they could not simply wander over it until they found a good place to settle down. Másaw, its guardian spirit, outlined the manner in which they were to make their migrations, how they were to recognize the place they were to settle permanently, and the way they were to live when they got there.

—Frank Waters, *Book of the Hopi*

"Where are you going?"
"We follow the cattle."

—Wodaabe nomad, breaking camp

Nashoba Brook 

At Butter Brook we find a convenient series of rocks and manage to get across still dry-shod. Then we climb a small hill, turn south again, and head for Nashoba Brook. We are going to have to cross another road soon, and we want to get as close as we can to it in deep woods, but we seem to be trapped here between wetland and gravel pits. At one point in the woods, in the middle of nowhere, we come across a strange series of blocks tied around the trunk of a tree. It takes us awhile to figure out that this is a deer stand. Somebody with a bow and arrow climbs up the tree before dawn and then waits in the tree for a passing herd.

As we inspect the deer stand Barkley wanders off in search of birds. The woods is still alive with little *chips* and *checks,* even though it is by now late morning. October is a serious time for bird life. From all over the Northern Hemisphere, from Michigan east to Nova Scotia—charged by the primordial memory of the warmer climates, where, according to one theory of migration, they lived in the geologic past—the great hordes of passerine birds move south in fits and starts. It is a convenient strategy for getting through the winter. It is also a strategy that, until very recently in geological time, was used by the human community. Until agriculture was invented, about eight thousand years ago, people didn't settle in one place; up until that time, nomadism was the standard way of life.

Nomads have produced some of the most fearsome people on earth, beginning with the horse-breeding Scythians, whom even the warlike Greeks feared, and including the Mongol hordes that swept over Europe in the thirteenth century and the Fulani and Tuareg tribes, which terrorized the inland explorers of north central Africa during the nineteenth century.

Nomadism is a dying way of life in our time. A few tribes of Gypsies, the Qashqa'i of Iran; the Tuareg, Fulani, Bororos, and Al Murrah bedouins of the north African drylands; plus a few remote hunter-gatherer tribes are the last people on earth for whom travel is not recreation but a way of life. These do not seem to be a people with a great loyalty to a single place. They abhor borders, pay allegiance to no governments, and are the bane of benevolent modernizers who would tame them and settle them into a comfortable (and controllable) agricultural existence—which is death for the nomad. Nomads are ruled by the hunt or the herd, game and the flocks determine the destination, and what little they own goes with them—tents, pots, sleeping mats, rugs, clothing. They weave their folktales and mythologies into movable artifacts, the elaborate oriental rugs of the Baluchi, the Bakhtiari, and the Qashqa'i of Iran, or the exquisite hammered jewelry, horse bridles, and patterned saddles of the ancient Scythians. No settled people envy them, except perhaps restless English wanderers such as Sir Richard Burton, Wilfred Thesiger, and the late Bruce Chatwin. And yet the concept of nomadism, the idea of the road, still has some atavistic draw, particularly here in

America, where the dream of the Western frontier held sway for so long and where a literary genre devoted to the road developed.

I once saw a big pack of motorcycles haul into a service station beside the big 495 interstate that slices across our pilgrim route. They were the type of middle-aged bikers from the mid-West who skin themselves in leather, pack more than they need into neat side boxes, and communicate with one another by means of radios, like fighter pilots, the microphones strapped to their helmets. This group was analyzing a map. They were deep in discussion, bent over the spread-out sheets like generals. One of them glanced up as I walked by.

"What the hell's the name of this place, anyway?" he asked.

It was a good question, I thought.

"Where're you headed?" I asked, after I told him the current name of the town.

"Canada."

"Where?"

"Don't know. Ain't never been. We'll decide when we get there."

Nomads, of course, move with direction. Unlike the wandering, lost Americans, wheeling along at sixty miles per hour just for the sake of the trip, nomadic people know exactly where they are going and exactly why they are in a certain place at a certain time. In fact, it could be argued that in spite of the fact that they rarely stay put for very long, there is no people with a more highly developed sense of place than the nomadic tribes; they read water-table levels by the quality of grass, they know every stone along their route, they navigate by the patterns of clouds, or by the lay of hills, or by watercourses dried for a hundred years. We are a lost people by contrast. We do not know our place.

Nearly every American family, given the resources, at some point plans and undertakes what one friend of mine terms "the great journey." Maps are laid out, a ton of equipment is purchased, trailers and camping gear are assembled, itineraries are plotted, and then some fine midsummer morning, the group sets out for the wild frontier, the Rockies, the national parks, Canada, Europe, or some nearby state park or city. Destinations vary. Some go at it in a serious way—quit their jobs, buy boats, take the children out of school, and sail away from the known world. Others never get beyond the limits of their own state.

In 1930, while he was working in a shipyard in New York City, Kata's father, who had not been far beyond the state of New York, signed on as deckhand on a three-masted schooner with the Whitney expedition to the South Seas, which had been organized by the American Museum of Natural History. They had a successful outbound passage and for months cruised from island to island collecting specimens and making sketches for exhibits, which, incidentally, are still on display at the museum in New York. At the beginning of the return voyage the vessel was wrecked on a reef. Kata's father escaped and, finding himself ashore without a ship, spent the next year in the Pacific. He and a friend had an idea that they would sail back on their own. They managed to purchase a forty-foot ketch and, after many ordeals and not a few storms, crossed the Pacific to San Francisco, where Kata's father remained for the rest of his life, building canoes in his backyard. Now, if you ask him to join you in an expedition to climb Mount Tamalpais (which is no more than a mile or two from his house), he will graciously decline. "Can't make it today," he'll say. "Got to stay home and feed the cats."

One needs go but once.

Some shouldn't go at all.

An old farmer I knew in Connecticut told me about a trip he and his family once undertook to drive all the way to the city of Worcester, a distance of approximately one hundred miles. They spent weeks planning the trip, and when the big day came, June 9, 1953, they packed maps, camera and film, emergency equipment, and a picnic lunch. While they were eating by the side of the road, they saw in the northeast, in the area where they were headed, a vast billowing thunderhead. This was no ordinary summer storm; the clouds rose in a vast dome thousands of feet, and even though they were fifty miles distant, they were clearly visible. When the family got to Worcester later that day, the place was in ruins. The great tornado of '53 had swept through the city.

The old man never left the farm again.

All this voluntary travel is luxury, of course, and is partly due to the democratically inspired right to travel and settle where you wish. In many countries you have to live where the state tells you to live. But here in America, mobility has worked its way into the culture. Whenever times get hard, we undertake to exercise our liberty to move. During the Depression and, more recently, during the era of Reaganomics, there was a segment of society that was on the road not for entertainment but for survival.

A few years ago, during the unpleasantness wrought by the questionable economic and social policies of the

1980s, I met a family in a park in Tucson, Arizona. The five of them lived in a Ford Maverick and a tent. They had a trailer where they kept all their worldly goods, and at night they would rig a plywood platform across the seats from the dashboard to the back window shelf where they would all sleep. The father, a man named Ray, had been a locomotive repairman in Pennsylvania. He had achieved the American dream, a house with a yard and an above-ground swimming pool, but when he got laid off from his repairman's job he couldn't find work, couldn't pay the mortgage, and since everyone else was in dire straits in that region, couldn't sell his house. The bank foreclosed, and so, fueled by the uniquely American myth of the frontier, the family hit the road. Never mind that the frontiers were closed.

"The kid here's got asthma," Ray told me, "so we come south. Supposed to be work in Texas, but we couldn't find any. Then a norther started to blow; kid got sick again and we had to put him in the hospital. When the norther quit he got better and we moved on. Came back east to Alabama, where we heard they were hiring in a yard there. No luck. So we went up to Oklahoma, no work there, like we'd been told, so we started west again."

His wife was sitting at a picnic table in the park with papers spread out before her, held down by rocks because of the wind. She was working out their budget. They still had some money and were planning their next move. But Tucson was all right with them for the time being. It was warm, and their boy was better. Over on the swing sets, three children, all under twelve, were swinging madly and leaping from the seats at the height of the arc. The smallest was a tousled-headed blond with a narrow chest.

"It's him that's got the asthma," Ray said.

My daughter, who was about the age of the youngest, was eyeing the wild troupe at the swings and edging closer, while we talked. Within five minutes she was up there with them, leaping from the swings as if she were part of their gang.

She and I were in Tucson on vacation (or research—I've never quite been able to distinguish "research" from whatever else one does in a given place). A few days earlier we had been at one of the pilgrimage sites of the Southwest, the exquisite little San Xavier del Bac mission just south of Tucson, known locally as the White Dove of the Desert. The church is situated on reservation land of the Tohono O'odham Indians, and on the day we were there, a festival was taking place. Part of the celebration involved a dance, a tedious, shuffling circle with long monotonous chanting and a steady, pounded, somber rhythm. The Indians were worried that their sacred mountain, Baboquivari, was about to be destroyed as a result of federal policies which would remove 2,065 acres on the eastern slopes from protection as a wilderness area, thus opening up the area for mineral extraction. Baboquivari was—still is, I suppose—the home of I'itoi, the elder brother, the spirit who created the Tohono O'odham. After finishing this work and after many adventures, I'itoi retired to a cave on the slopes of Baboquivari, and the place came to represent the center of the earth for the people, the first place.

Before the dance, one of the Tohono O'odham made a speech about the issue of the destruction of the mountain. (They had also offered a few prayers at a mass held earlier in the day—not bothered, apparently, by any apparent conflict between the Catholic doctrine and their indigenous spiritual beliefs.) There were a few white-belted tourists in cowboy hats standing around the dance, eyeing it uncomfortably, but because the dance was boring and continued on for what in Western terms was an inordinate length of time, most of the watchers wandered off to buy fry bread and trinkets at a nearby market. The ceremony went on without them—a great shuffling circle, the drum throbbing, the men chanting in high, slurred vocables. I had just been reading Vincent Scully's theories on Native American dance and its relationship to landscape, and although I couldn't understand the words, I began to get the point. They were making peaks and valleys, cactus and coyotes, and arroyos and flatlands; they were recreating landscape.

The architecture critic Vincent Scully says that the Greek temple is the sculptural expression of the character of a god in a specific place; landscape, he says, had sacred connotations for the Greeks, and the temple and the site and the god cannot be separated. Some years later he said the same thing about Native American dance, that dance is at once a reflection and a creation of the landscape in which it occurs. In fact, he believed that Indian dance was the greatest art form ever produced on the American continent.

The Tohono O'odham dancers were telling a story in that tradition, the story of I'itoi and how he came out of the mountain and created the people and how he fought his enemy Siwani, and how I'itoi was killed by Siwani and lay in the mountain for a while and then came alive again and how, after many years, he went back to the mountain. The mountain was Baboquivari and that mountain is the people and the people are the mountain and if you destroy Baboquivari you will destroy I'itoi and if you destroy I'itoi you will kill the Tohono O'odham.

This is a tale that has been sung by others. Long ago, in Dreamtime, the spiritual ancestors of the Australian Aboriginals sang this same song. They created the world by singing it into existence. The lines of their songs became landscape, and even today, if you know the melody, the songline, you have the equivalent of a map of the world so that in theory you could find your way across the continent simply by singing. Some years back the Australian government undertook to lay out a new rail line that—for once—would not intersect or destroy an aboriginal sacred site. After much work and research and consultation with elders, no route could be found: the whole continent was holy.

Later in history, after the invention of agriculture and writing, many of these folkloric tales were written down and reappeared as epics—the story of Gilgamesh, of Noah and the Ark, the Ramayana, or the Odyssey. These are not just folktales, they are metaphorical histories of the land, and they often describe the relationship of a given culture to its environment.

While we were hiking along the high ground between Butter Brook and Nashoba Brook—silent for once—it occurred to me that somewhere in all this there must be the environmental equivalent of the grand unifying theory of physics—the great underpinning of the universe that Einstein spent the latter half of his life searching for and that modern physicists jokingly refer to as the Holy Grail. It struck me that there must be some cultural counterpart, some tie that binds together the finely crafted Scythian animal effigies of the first century B.C., the nomadic twentieth-century Americans out on the road for no purpose, the animal, floral, and folkloric patterns of the Qashqa'i rugs, the siting of Greek temples, the creation of the world in the American Southwest by a being who lives in a cave on a certain mountain, the founding of Concord, the search for the Fountain of Youth, the search for the Holy Grail, the number of hobos riding the rails during the Depression, symbolic dances and songs recreating land and myth, and wild bands of nomadic Tuaregs murdering exploring Europeans at the water holes of northwest Africa. Walking there towards Walden Pond, through a landscape layered with human histories, it seemed to me that the one commonality among these disparate events and artifacts has something to do with the human relationship to place, to *querencia* and the Hopi idea of *túwanasaapi.*

Nashoba Brook has been collecting waters all along its route as it meanders towards the Concord River. The streambed has been widening and deepening along the way and like the other small brooks it is proving to be our nemesis. We keep casting about along its boggy banks in search of a crossing. In one wet area, poor Barkley goes ankle-deep in mud, his new boots squishing and caked. We have to retreat to high ground above the marshes while he empties and cleans them. There are some white pines on the south-facing slope, the area is clear of shrubbery, and the needles have made a soft bed; we lie back while Barkley unlaces, and watch the autumn wind lilt through the upper branches. The earth smells of warmth and pine and dry, distant summers.

This is proving to be a much longer trip than we expected. Once again we are hiking in the wrong direction in order to stay dry-shod. We might be better off just wading through marshes and ponds and streams and making a direct line, but over the years we have become fastidious.

Once Barkley's boots are cleaned we hike on and eventually come to an old bridge on a gravel road that crosses Nashoba Brook. It is a pleasant site, an arched stone construction with the brook pouring beneath it and, slightly downstream, a pool with a small waterfall above it. We stop again to admire the spot and reconnoiter. South of us, on our direct route, is an area of gravel pits, horse trails, and small, isolated industries. But if we head north-northeast, away from Concord (again), we can travel along a wooded ridge just south of the Carlisle State Forest and then trek southward along a series of ridges to the next major barrier—Spencer Brook.

While we are sitting on the bridge debating our route, we hear the clip-clop of horse hooves, and a pony cart wheels around the wooded bend at full speed. The small, delicate cart is painted red and blue, with bright yellow spokes on the two wheels. An older, well-muffled couple in tweeds and shawls and fashionable peaked caps and a tartan blanket spread across their laps sit primly side by side, the man at the reins.

Barkley greets them as they pass, and the little pony reins in with much snorting and stamping.

The driver, a florid-faced man in his early sixties, asks whether we are lost—he has spotted our maps.

"Lost in debate," Kata says, as we explain our dilemma.

"Well, there's a beautiful road through the swamps and trees just beyond this bridge," the driver says, "but the nicer part is west, wrong direction for you. We're headed there now."

"A fine day for a pony cart ride," Kata says.

"He's a spirited little beastie," the woman says.

In fact he's quite big for a pony, more like a small horse, and he has a white star blaze on his forehead and a shiny chestnut coat. While we talk he raises and lowers his head and snorts periodically, as if he's anxious to be off. His owners are in no great rush, however. They seem interested in our quest.

"Why would you want to *walk* to Concord, when you can ride there?" the man asks.

I notice that Kata shoots a knowing glance at Barkley, but the man means ride in a pony cart to Concord—in style, that is—not in a boring car.

"You can ride all the way in a pony cart?" I ask.

"Oh sure. Or at least most of the way. Give or take a road or two. And the hunt clubs range all over these parts, from Lincoln to Westford. It's a fine sight when they're out, the dogs, the riders in traditional garb. You must go through the Estabrook to the bridge. There's a good road for pony carts there, many horses. That's the road the Carlisle Minutemen marched down to the Concord fight."

Since he seems to know his local history, I ask him about the Westford contingent.

"Well, John Robinson went down on horseback, ahead of his men," he says, "so he probably took the Great Road most of the way, plus a few short cuts. But the rest of them marched down the old Acton Road, just east of here. You'll have to cross their line of march to get to Concord from here."

"The Concord Convergence," Barkley says.

The pony-cart people miss the drift, but I know what he means. From Westford and Carlisle, from Littleton and Bedford and Sudbury and Stow and Groton and Chelmsford, all the minutemen and militias were converging on this singular place, this obscure wooden bridge above an obscure tributary of the Merrimack River on the Eastern Seaboard of the Americas, and the only thing separating us from them and them from us is this strange Western concept of time.

"What a glorious day," Kata says.

"That it is," says the charioteer, and snaps the reins. The little horse leaps forward and they spirit off around the next bend and head west, up the gravel road.

"Where's Bradley?" Rogers asked.

"Major," said Kelly, a red-headed Irishman from Suncook, "he went home."

"Home!" Rogers cried. "What you talking about?"

"Major, he said the Cohase Intervales was just two days from his home in Concord, and the quickest way to get back was head straight for Concord. He said all of us could have supper at his father's house day after tomorrow if we went that way."

"Concord!" Rogers said. "Where in God's name does he think Concord is?"

—Kenneth Roberts, *Northwest Passage*

North Street 

SINCE WE ARE NOT GOING TO BE ABLE to make a direct line to the bridge and since the area to the southwest is hopelessly developed, we head northwest away from Concord and soon approach another road, Route 27, which runs through Acton. Once more we are crossing paths with the Westford Minutemen; they would have been tramping along this road sometime after sun-up, tiring no doubt by this time, but still up for a good fight. Route 27 is not heavily traveled, but it's jarring nonetheless after our woodland, swampland, marshland sojourn. Furthermore, to our right as we approach the road we can hear more gunfire. One would think, given the amount of firearms we have encountered this day, that the country was preparing once again for war. This time the noise is coming from the Nashoba Sportsmen's Club, where, on this national holiday, our attentive compatriots are presumably celebrating the arrival of Columbus by sharpening their shooting skills. All this target practice and gun toting here in thc thoroughly modernized, relatively safe, nonhunting hinterland of eastern New England seems out of place. But the gun is very much a part of the American mythos of the frontier, and perhaps it endures as a sort of antithesis to rootedness, a symbolic rejection of place.

The Westford Minutemen were heavily armed with muskets and swords, powder, and shot as they marched along the road in front of us. Most of the men owned their weapons, and those who didn't had them supplied from town munitions. Some of the guns were ancient pieces not designed necessarily for war, long-barreled fowling pieces, Spanish fusees, French muskets and the like. Shot and powder were precious and would have been distributed from town armories, some of which were actually located in churches, which says something about the nature of the upcoming struggle. The men would have carried their bullets wrapped in handkerchiefs or stored beneath their wide-brimmed floppy hats. It was chill that morning, so many were dressed in waistcoats and coats of varying color. They had long hair, tied back; they were cleanshaven; they marched in ordinary cowhide shoes with wide buckles and long stockings of homespun. The colors were those of the April woods around them, the gray and brown of dyes made from local oaks and sumacs and hickories. They may have seemed a benign, somewhat disorganized collection of good farm boys, but appearances were deceiving, even to the British. These men were well trained and disciplined. Inevitability was in the air that day. In spite of the fact that the battle at the bridge was the result of a misunderstanding that might have ended in a negotiated settlement, the presence of arms and soldiers, and years of training, determined the nature of the outcome. America was already at the mercy of its weaponry.

Mixed with the sound of gunfire from the Nashoba Sportsmen's Club is the occasional pounding of a passing truck, and soon we break out of the woods and stand in awe beside the road. Litter, of course, is the first thing one notices on these occasions—beer cans, candy wrappers, and cigarette packs; one wonders what the roadsides would be like if people didn't drink beer from cans, smoke cigarettes, and eat wrapped candy.

For a few hundred yards we hike along the road and then turn off onto a small side street that will lead to the high ground to the east. But, as we are slowly coming to realize, our map is either sadly out of date or inaccurate. The street soon breaks into a series of smaller roads of a neighborhood that does not appear on the map. Hiking bravely past barking dogs, curious children, and suspicious elders in their front yards, we move in what we take to be the right direction to get us out of this warren of streets to the wooded ridge.

The houses in this neighborhood are small and well taken care of. Some of them have white picket fences with little flower beds in front. Below the windows and around the doors of each house there are yews, junipers, and rhododendrons, the requisite foundation plantings so beloved of the good people of this country. Attached-garage doors are opened in some of the dwellings and inside we can see a vast array of equipment employed by the natives in the care and feeding of their children, dogs, and yards—lawnmowers, snowblowers, lines of rakes, stored baby carriages, playpens, plastic swimming pools, bags of dog food, and carpentry tools. The lawns around the houses are green and well tended—some of the males are out raking away the fallen leaves, and some are spreading fertilizers and lime. Not all the good burghers are suspicious of us pilgrims. Some wave as we go by, and some of the dogs charge out barking, as if to attack, and then, sensing perhaps the onset of a fine expedition, a good hunt, join with us as we walk. In a few minutes we are hiking along with a motley crew of three or four of them.

There is a small pack of young children on a swing set in one yard, and one of them shouts something to us as we go by. We wave back and they come over to the fence and pronounce a name that we can't quite catch and then break into gales of laughter.

"I think it's a TV personality," Kata says.

One of the braver ones, a small girl of about six or seven in pigtails and a red sweater, asks us where we're going.

"We're off to see the wizard," Kata says.

She seems to draw a blank.

"No you're not," says an older boy in the back of the pack.

"Oh, but we are" says Kata.

"We are indeed," says Barkley.

"No you're not," says the boy, "BECAUSE THE WIZARD'S NOT REAL."

They all break into peals of laughter again.

Dorothy would have understood this neighborhood, epic hero that she was. She had learned about the sense of place the hard way. Guided by the traditional wise fools and tricksters of folklore, and accompanied by the loyal companion Toto, she sets out for the netherworld to seek the boon that all heroes must find. Through her wile, she overcomes a series of obstacles, smashes the illusions of the mysteries, and gains her boon, which she carries back with her to Kansas in the form of a lesson. Her power to return, as she learns from one of her spiritual guides, the good witch of the north, is within her, and was all along, it's just that she had to learn about it for herself. "If I ever go looking for my heart's desire again," Dorothy says, "I won't look any further than my own backyard, because if it isn't there, I never really lost it to begin with." In order to get back to Kansas, she has but to repeat a magical incantation: "There's no place like home."

Dorothy's story, the neighborhood, the gridwork of streets, are all a part of the American vision of place—a little house with a yard and a white picket fence, a good home with two cats in the yard. A survey done a few years back by the National Bank Board revealed that the dream of the single-family home still endures in America and that the ideal is especially keen among Hispanics and African Americans. It was an image formed somewhere in an imaginary England, where all the world was green and everyman's home was his castle, and for a few it came close to realization in America in the late nineteenth century, when the small village consisting of single-family houses reached its apotheosis in this country.

It's a dream that has extracted a heavy toll on the land. The Harvard biologist E. O. Wilson has postulated that human beings are most contented when they are located in sparsely treed heights, overlooking water. He says such landscapes recreate the ancient savannahs of Africa, where the human consciousness was nurtured and where the first dim concepts of an ideal landscape developed. Wilson points out that whenever people of power get control of the land, the first thing they will do is construct a simulacrum of the green, sparsely treed height—whether chateau, castle, or corporate center. The result here in America is the suburban house, with its lawn and trees.

Here in the well-watered East, where the forest is lurking at every yard edge and where the concept of natural landscaping has taken hold, the re-creation of this ideal has not exacted as exorbitant a price as it has elsewhere in the country. If you fly over the New England suburbs in a small plane when the leaves are out, you seem to be sweeping across one great forest—never mind that that forest does not harbor the plants and animals that it once did. In other parts of the country, though, the great sprawl of suburbia has supplanted the native vegetation to such an extent that it can never return. And in dry areas, where the atavistic memory of green grass beneath trees seems especially strong, some of the greatest stresses on limited water supplies are caused not by drinking water and sewage disposal but by lawn watering. In cities such as Tucson, which rely on ancient strata of unreplenishable groundwater, the practice is especially damaging. In fact, water is the limiting resource, and like the dearth of firewood in more ancient cultures, lack of water may eventually doom the population centers of the American West to extinction.

After a series of rights and lefts and a few wrong turns, we come to a street that runs along the base of the wooded hill. The houses are fewer at this point and set in the deep shade of large pines and oaks. One house stands out from the others. It is a small, unkempt, and nondescript place with plastic sheets nailed over the windows, closed-up doors, and a litter-strewn yard. An unattached garage with an open, broken door is filled floor to ceiling with newspapers, old windows, plastic garbage bags, tires, and rusting lawnmowers. A winding brick path, much spalled and rucked, leads to the backdoor. Inside we see a single light shining in the gloom.

Barkley and Kata halt abruptly and eye the place unabashedly.

"Are we sure we know where we are?" Barkley asks me.

"Well, not entirely," I say. "I thought we were just south of Rail Tree Hill."

"Shouldn't we ask?" Kata says.

Before I can say it is not absolutely necessary, and not at this place in particular even if we were to ask, they are striding up the path to the backdoor. Kata knocks, but there is no response. Then just as we are about to leave, the door is ripped open. A great billow of cigarette smoke pours out and a man of about thirty-five with curly black hair, a narrow face, and heavy black plastic glasses asks what we want. Behind him a much-lined woman, who could be either his sister, his wife, or his mother, appears. We explain our mission and he joins us in the yard. His female partner stays on the small porch, smoking.

He does not seem at all baffled by our pilgrimage. In fact, he seems to indicate it's a good thing to do.

"You climb this hill here," he explains. "You'll come to a pipeline. Tennessee Gas Company. Okay? Then you turn, head down the line, you'll come to an old road. Real nice in there. A lot of deer too. I've seen them myself. Matter of fact, I saw some the other day. I wouldn't kill any though. Matter of fact, I'm not a hunter. At the road you turn east, keep going, okay? Might see some logging in there, cut trees, but keep on. You'll come to West Street."

He rambles on in a fast pace about the hill and its denizens. Then he begins to talk about himself.

"Know what I am?" he asks.

"What are you?" asks Barkley.

"Jack of all trades," he states, "master of none. That's me. On disability now on account of my back. I hurt it on the job. But I know a lot of trades. I can fix anything."

His father had made some money in the excavating business. When he was no more than fourteen, he tells us, he himself had helped drive the machines to create one of the shopping centers over on the Great Road. But for reasons he does not explain, he quit working for his father. He washed dishes for a while at one of the Great Road restaurants and then, so he claims, because he was so good with machines, was elevated to the position of night janitor. His story grew faster and more elaborate, with many prologues and many rambling asides. It built to a climax on a snowy night the winter before, when he rose from his bed, alerted by a dream, and went to the restaurant to fix the furnaces. He knew by his dream something had gone wrong. That night, in the furnace room, he slipped, or lifted something, and his back was forever destroyed.

"Can't work now. Back's no good. I got the disability."

"So what do you do now?" Kata asks.

"Collect eagles," he says. "Matter of fact, I got a thousand eagles. I scour the fleamarkets every weekend. I got statues, belt buckles, pins, paintings, badges, emblems, you name it. You want to see?"

I can see Kata's interest building.

"We're probably going to have to move on if we're going to get to Concord by nightfall," I say.

When it comes to people and conversation, Kata tends to lose track of time. She can stay on her feet for hours listening to long pointless stories, and has the ability to read deep metaphors into simple accountings of dogs, or children's behavior, or the perversity of cousins. Sometimes her lack of guile gets her into trouble. Men, especially exotic men, often misread her interest and invite her home to bed. Once, in the isolated Berkshire town where she lives, she met what must have been the only Senegalese man in all of Connecticut. Because he recognized a set of Hopi rabbit sticks that she was carrying, she took him for a mystic or a messenger. He had other intentions, though, and vowed he would divorce his other four wives if only she would marry him and come back to Senegal.

"Well, there's a lot of animals up there on that hill where you're going," our guide says. "I can tell you all about them. Indian ruins too. Burial ground."

"On our route?" I ask.

He gives us directions to the purported ruins and we try to ease away, but he follows us down the street in order to show us a small trail that mounts the hill.

"Matter of fact," he says, apropos of nothing, "I *like* raccoons. I raised ten of them. Then I let them go right on this hill. Some's tame and come back. Neighbors don't like it though."

We thank him and bid him farewell and begin to climb the steep hill. He remains behind, standing with one hand on an old stone gatepost, shouting directions up to us. As we disappear into the trees, he moves up the hill.

"You'll come to the pipeline soon," he shouts up at us. "You just turn there."

"We will," Barkley calls. "Take care of your back."

"I will. Don't get lost in there."

"We won't."

"If you see some raccoons, it's probably one of mine."

"We'll say hello," Barkley calls back.

"Don't forget to turn at that pipeline."

Jacob then awaked out of his sleep, and he said, Surely the Lord is in this place; and I knew it not. And he was afraid, and said, How dreadful is this place; This is none other but the house of God, and this is the Gate of Heaven.

—Genesis 28:10–14

To enter the sacred space is therefore to make a pilgrimage to the center of existence, and for this reason sacred spaces are typically conceived as centers of universal community.

—T. Ronald Engel, *Renewing the Bond of Mankind and Nature*

Nagog

As we suspected, the pipeline proves to be a hideous swathe of treeless land overgrown with running blackberry and brambles. The pipe itself is underground. Furthermore, although the route suggested by our guide is more direct, we are trying to avoid streets and houses by walking north, around them. We do make a point of visiting his "Indian burial ground" though.

As we are crossing the line, we hear a great snarling and roaring in the distance, and soon, over the brow of the hill, a smoky company of armored knights appears, riding brightly painted trail bikes. There is no telling what age, sex, or nation these knights belong to. They are covered from head to toe in shiny armor, complete with the gauntlets, breastplates, jambs, helmets, and cuisses of the traditional knights of old, except that this armor is fashioned from high-tech industrial material. Furthermore they are coming on fast, surrounded by billowing clouds of dust, branches, twigs, the brush parting as they charge. Once again, were we pilgrims of the Middle Ages we would at this time be experiencing a deep fear for our very lives. And in fact we must be showing some level of trepidation; as they pass, one of them pulls over in a cloud of dust and leaves, and leans forward over his handlebars. "You guys all right?" he asks, without lifting the black visor of his helmet.

We assure him that all is well, and with a twist he turns his mechanical steed and spins off to join the others.

Halfway up the hill the forested land shows signs of having been cut over. We come across wide stumps and young growth everywhere, and once we are off the beaten track—which runs out after a few hundred yards—the walking is difficult. Hawthorn, young oak trees, brushy slash piles, and a thick understory of high-bush blueberry force us to pick and weave through the few larger trees that were left standing. North of this area is a natural uncut woods and here, on uncharacteristically flat ground, we come across a series of stone mounds, each about eight feet long and approximately three or four feet high.

"The famous North Acton burial grounds," Barkley announces. Barkley does not believe in Indian burial grounds, at least not stone mounds.

The land over which we have been walking today is interlaced with stone walls. They run up the sides of hills, they have provided convenient passages across wet meadows and marshes. We saw them in the deepest woods, in light woods, in brushy pastures, along roads, and, of course, they line every meadow and field we walked through. Half the stones in these walls are cumbersome weighty things that can not be carried by hand. They were loaded onto stone boats, hauled by oxen to the property lines or field edges, and then unloaded and stacked to a height or four feet or so. Not long ago, these walls were universally believed to be the work of Yankee farmers intent on clearing their fields. But in recent years, a theory has developed that some of them were the work of the local Native Americans.

One of the foremost proponents of this theory was a man named Byron Dix, known to his detractors as Lord Byron. Some years back, Dix wrote a book in collaboration with a man named James Mavor, who was one of the American scientists who helped unravel the mysteries associated with one of the great mythic places of the Western world, Atlantis. Mavor and his Greek associates developed the now generally accepted theory that Atlantis was located on Thera in the Cyclades in the Aegean.

Both Dix and Mavor had scientific backgrounds, although they were better versed in engineering than in archaeology. The book they published, *Manitou: Sacred Landscape of New England's Native Civilization,* propounds, in highly scientific language and with an engineer's eye for surveys and statistics, the theory that many of the stone walls of New England were constructed according to the solstices, equinoxes, and the rise and set of certain stars, not unlike Stonehenge and other archaeo-astronomical sites. Lord Byron and his followers located, sometimes in vast configurations covering hundreds of miles, mounds of stones that bear no apparent relationship to past agricultural practices or property lines. These stone piles, the so-called Indian stone rows—which are larger and messier than the Yankee walls—and other stone constructions spread out over New England were lined up not only according to the stars, Dix and Mavor proposed, but also along lay lines, or lines of spiritual force that crisscross the earth, intersecting at certain spots. These spots, according to the theory, are recognized sacred sites: the pyramids and temples of Egypt and Mesoamerica, the stone circles and cathedrals of Europe, the mounds of the Hopewell and Adena cultures of the Miami River Valley in Ohio, and now, perhaps, the stone rows, mounds, and standing stones of New England and Canada.

Our mounds are placed in orderly rows, and are similar to some of the barrows I've seen in England near Silbury Hill and Avebury. Except for the lack of slope, the land around the mounds is undistinguished, with a few older oaks, a sizable sugar maple, and two American beeches—actually, an odd assemblage of trees for the region. No doubt the area was cultivated at some point. From this spot in winter you would be able to see out to the western hills of Nashoba and Newtown Hill, but it's not a particularly significant view, and nothing suggests that the site could line up with the setting sun, or the rise or set of significant stars or constellations. The mounds run north and south, roughly, and there doesn't seem to be a pattern to them. According to our way of thinking, mine at least, and Barkley's, these mounds were probably built by sheep farmers clearing pastureland sometime during the nineteenth century when Merino sheep were all the craze in the New England farming community and everyone was giving up on wheat and barley.

Kata, who is always looking for mythic possibilities, is more open-minded.

"Is it not possible," she says pointedly, "that you just don't know how to read the stones or what their purpose was."

At least Lord Byron's stone rows are theoretically the handiwork of people who were actually known to be living here and elsewhere in the Americas; furthermore, Indians actually did construct archaeo-astronomical structures. There have been more outlandish theories. Here in the United States, ever since the mid-nineteenth century, amateur archaeologists have detected all manner of pre-Columbian visitations and colonies. One of the least grandiose stories is our legend of Sir Henry Sinclair and the Westford knight. But others were in vogue before him: the great Viking tower at Newport, Rhode Island, which is probably the work of sixteenth-century Portuguese; the skeleton in armor, the inspiration for Longfellow's poem of the same name, which proved to be the remains of a Native American with a bit of brasswork on his chest; the Kensington Stone, the purported remnant of a thirteenth-century Viking expedition in Wisconsin, which proved to be a hoax; and more recently, theories of Irish Culdee Monk cells in New Hampshire and Massachusetts, Phoenician and Iberian hieroglyphics on standing stones and monoliths in Vermont, runes on Dighton Rock in Buzzard's Bay, runes on Manajna Island off Monhegan Island in Maine, runes on Noman's Land off Martha's Vineyard, runes on every rock and boulder, and all of them considered by traditional archaeologists to be highly suspect.

Stories of exotic or spiritual visitations in certain select places on earth are not confined to the New World, of course. Early Christian saints seem to have possessed a remarkable ability to perform miraculous voyages. A few years after Christ's death, St. Thomas seems to have made a voyage to south India and founded a Christian sect, which endures to this day. St. James traveled to Iberia and converted the natives, but made the mistake of returning to Jerusalem, where he was beheaded by King Herod. After his death his disciples transported him back to an obscure site on Cape Finisterre in northwestern Spain and buried him there in the territory under control of Lupe, the pagan queen of wolves. Nine hundred years later, guided by a shower of falling stars, Charlemagne discovered his body, unscathed by death, and built a sanctuary there. The place evolved into Santiago de Compostela, one of the most important pilgrimage sites in Europe.

Shortly after the death of Christ, Mary, Martha, and Joseph, accompanied in some versions by Mary Magdalene, landed on the coast of the Camargue in the South of France. The spot, Les Saintes-Maries-de-la-Mer, is now an important pilgrimage site for Gypsies. The chapel there contains a bejeweled statue, one version of the Black Madonna, clothed in rich vestments and venerated as Santa Sarah by the Gypsies. The town is a vacation site and, according to one description, the saddest place on earth—until the Gypsies arrive for their annual festival.

Somehow, Joseph of Arimithea made his way to England, carrying with him the *san graal,* the sacred cup from which Christ drank at the Last Supper, thus inspiring fourteen hundred years of pilgrimage and expeditions in search of the lost Grail, one of which may have been the voyage of Sir Henry Sinclair to Prospect Hill in Westford, Massachusetts.

In our time, one million Poles descend on the monastery on Jasna Góra hill each August to witness the unveiling of the image of the Black Madonna, a portrait thought by some to have been painted by no less a figure than St. Luke himself. Periodically the portrait sheds live tears, and once it bled when slashed by a fanatical soldier.

Miraculous transformations of places into sacred sites continue to occur. Some years back, by the sort of surreal alchemy that seems to take place on the Indian subcontinent, a temple in a distant forest of the Sabari Hills in the Western Ghats of Kerala was somehow determined to be the actual site in which Lord Ayyappa, a primal forest deity who was adopted into the Hindu cosmology, had his dwelling. According to the folklore, after fifteen years' banishment in the forest, Lord Ayyappa came riding out to civilization on the back of a tigress. By means of government support, the site of his palace has now evolved into a pilgrimage place and some eight million people hike the twelve miles through the forest each year in January during the Lord Ayyappa festival to reach the temple. Adherents must abstain from sex for two months before they make the journey and women of childbearing years are forbidden altogether.

None of these miraculous saints' voyages (save perhaps that of St. Thomas to India) nor the sacred role of any of the sites are based on historical fact. The roots of such associations are deeper than mere history. The Black Madonna in particular is associated with elemental earthly forces and ancient goddess cults. Her image appears in different sites throughout Europe, and she occurs also in India and Tibet in the form of Tara, a black-faced, powerful deity who is a sort of Buddhist alter-ego of the fearsome Hindu goddess, Kali, she of the necklace of skulls, who breathes fire and has a lashing red tongue and hideous popping eyes—the great destroyer, the great creator.

If you exclude Concord and the pilgrimages to Walden Pond, a few Catholic sites such as St. Anne's in Quebec, and the pilgrimage site at Chimayó in New Mexico, where the local Indians turned their sacred springs into a sanctuary to St. James (a New World

Santiago de Compostela), the idea of pilgrimage to holy ground in America has not fared well. The white Anglo-Saxon Protestants who forged the mores of this country were not given to taking up long, ecstatic journeys to spiritual centers.

Unlike the shrewd early Christian bishops in Europe, who turned the pagan deities into saints and built cathedrals on heathen sacred sites, the English Puritans preferred to destroy all evidence of the religions of the local godless heathen. The Native Americans' myths were ignored and their religions forbidden, and any sacred structures they may have built were obliterated by agriculture.

According to the native people, the land of America was invested with the Chinese equivalent of *chi,* or *feng shui,* or spirit power, before the Europeans arrived. (It still is, according to some Native Americans.) For the Wampanoags, who lived just to the south of Concord, this spiritual power was called *manit*. It would manifest itself in animals, in landscape, or even in the behavior of people. Around Concord, one of the personifications of this spiritual power was the being Hobbamock, who could take many forms and, as is often the case with powerful spirits, could destroy you or save you. Hobbamock and his compatriots Marshopa, Glooscap, and others appear in myths and legends up and down the coast, and they sometimes express themselves in landscape. One of the origins of the name Nashoba, which was used by the English to describe the mixed group of Massachusetts and Pawtucket Indians in the mid-seventeenth century, is thought to be Marshopa, or Moshup, a local spirit power known to the Wampanoags. Among other things Marshopa could ride on the backs of whales. It was he who created the colorful cliffs at Gay Head on Martha's Vineyard, Devil's Bridge and Noman's Land.

Just across the valley from the burial ground we are visiting, there is a low hill above a pond known as Lake Nagog. During the colonial period, that hill used to thunder or give off loud booming sounds like cannon fire. According to local Pawtucket legends, the four winds were pent up in the hill and periodically thundered and boomed in an attempt to escape.

In precolonial times, the view from the spot where we are standing was probably not different from the current aspect of the place, the surround of trees and shrubs, and a flat shelf of land with the slope of the hill to the east. Good vistas are hard to come by in this region, but even without a vista, this appears to be as pleasant a place as any to have a picnic, perhaps even propitious if the ground is indeed sacred, and so we decide to eat here. We clear a place among the stones and Barkley unloads and carefully unwraps his various delicacies.

Barkley is a man who delights in taking care of things. He now extracts his Swiss Army knife and uses it to open our bottle of Sancerre. That done, he switches blades and cuts off slices of a smoky Jarlsberg cheese, and then hunks of bread, which he passes around with ceremonial formality. If need be he could use this same tool to extract splinters, clean nails, pick teeth, drive screws, and even unscrew small nuts, but we have no need of that at the moment. We pass around the glasses, pour off a little wine and toast the holiday, Columbus Day.

"To the voyage of Cristobal Colombo," Barkley spouts. (He's being facetious; he and Kata do not like Christopher Columbus.)

Barkley sprinkles a little wine on one of the stone piles. "Forgive us his transgressions," he says.

"Don't fool around, Barkley," Kata says. "Some people would say you'll get yourself in trouble that way. Especially here, in this place."

It is true that elsewhere in the world, including the present-day American Southwest, certain sites traditionally have to be approached with caution—voices must be lowered, shoes kicked off, offerings or sacrifices left. The spirits that inhabited the land around Concord also had to be propitiated, although the most the Puritans ever say about it is that the Indians practiced a sort of devil worship. John Eliot, organizer of the nearby Christian Indian Village, Nashobah Plantation, which was located somewhere just to the northwest of Nagog Pond, tried his best to stamp out the native religions, but remnants have endured for nearly three hundred years, and are now being restored by the *nouveau* Indians, of whom, one might argue, Kata is a member. Eliot once preached a sermon to the Wamesett tribe from a large rock in Connecticut, now known as Pulpit Rock, which may have been one of these sacred sites. It's possible that he did this on purpose, to drive out the local spirits. If that wasn't his purpose, there is a certain irony in his choice. The service that day must have fortified the native view of the Christian God as just one more powerful spirit in their own pantheon.

In the early nineteenth century, rural white people in New England told stories of certain sites they called "sacrifice rocks," located throughout wooded areas that had yet to be cleared. One English traveler of the period, Edward Kendall, described these as stone masses standing in different parts of the woods and claimed that the remaining Indians, although long since Christianized, still made offerings on them. Kendall questioned some local Wampanoag women about the rocks, but they only would say that they had been taught to make the offerings by their ancestors. They weren't sure what would happen if they stopped.

One of these so-called sacrifice rocks stood on top of Nashoba Hill just to the northwest of us. Another may have been located around Egg Rock, at the confluence of the Sudbury and Assabet rivers, which combine at that point to form the sacred Concord. Thoreau, who was a student of local Indian customs, had the right idea about the local rivers and this site; he said that Concord's true emblem, its shield, should be a field of green with the Concord River circling nine times round.

"We have some potato salad," Barkley says. "Also sardines, a Vidalia onion, and my sulphur shelf mushrooms."

Kata unwraps her cold chicken *arachide* and distributes portions. She has also brought along a container of wild prickly pears, which she put up herself.

Barkley fumbles in his pack and carefully extracts the eggs mimosa. Then he rummages through Kata's pack and draws out a bar of Suchard chocolate, of which Kata is a known addict. They have grapes, and pears, and four local Macoun apples, purchased the day before from a nearby cider mill, and also a hunk of goat cheese from a Berkshire farm.

We are leaning against one of the cairns—or tombstones—our legs stretched out to the west, and we pass the food around and talk casually, without looking at one another. Above us the sky still holds that brilliant

October blue, which, when framed by the tinged autumn leaves, seems to have a uniquely American cast. All regions have their skies, their qualities of light; sometimes they can be powerfully affecting. One American pioneer I heard of never could be happy in his various Western homesteads. What he missed most was the light above his Vermont farm, a light he never found again.

We clink our glasses and toast again to *bons voyages.* We eat the eggs mimosa, toast once more, break into the cheeses, have more bread, then the pickled cactus and the chicken, then the apples and goat cheese.

Apples and pears are a specialty of the Nashoba Valley. This whole region was settled by people from Kent who must certainly have had some familiarity with orchards, since that county is famous for fruit-growing. One of the first things the settlers planted after they staked out farms here in the valley were apple trees, although they had cider in mind, not fruit. Periodically, as we have hiked through the higher grounds here, we have come across ancient apple trees among the oaks and white pines, still pushing out a few yet vital green branches from slanted, hollow trunks.

We could easily spend the rest of the day here leaning against these stones, eating, drinking wine, dozing in the sun, and talking of the transgressions of Columbus. Barkley is trying to argue some bizarre theory that the kiss of death for any culture is to be colonized by the relatively benign and enlightened English, as opposed to the brutal Spanish, or the indifferent, amoral French. Kata is not taking it well. "They were all murderers," she says.

I am watching the sky while they argue. A discussion such as this could go on for an hour. I know there will be future diversions, and I want to keep moving, but Barkley is presenting his closing argument, asking Kata to list extant tribes in North and South America and then asking her to name their "oppressors." It's true none of the surviving tribes she names were under the domination of the English, and I can't think of any myself.

Not long ago there was much talk of Columbus in liberal-minded Concord and its environs. One of Concord's adopted sons, a man named Fox Tree, had been working for a number of years to have Columbus tried as a war criminal. Fox Tree is of uncertain origin, but looks for all the world like one of Melville's chief harpooners, Queequeg or Tastego, with long black locks, high cheekbones, dark skin, and shiny black eyes. He claims to be Arawak, a Caribbean tribe that was presumably exterminated by the Spanish early in American history. He claims that Columbus was more evil, more guilty of crimes against humanity, than some of the darkest figures in human history—the Mongol Ghengis Khan, Tamerlane, even Hitler. During the quincentennial of Columbus's "discovery" of the Americas, Fox Tree was often a featured speaker in local events and prepared murals and posters depicting the many crimes of Columbus.

One of the chief henchmen of Columbus was our guide through Florida, Juan Ponce de León. He had come over to Hispaniola on the second voyage as a pensioned soldier, served with the brutal Indian killer Ovanda, and was associated with some of the most murderous expeditions in the Caribbean. He was so successful at killing Indians that he was granted permission to sail on an expedition of his own to the as yet unexplored island of Puerto Rico, where there was said to be much gold and a docile people waiting to be enslaved. Juan allied himself with the island's chief, killed or enslaved those who did not submit to his rule, and by way of reward was made governor in 1509.

Ruinous Spanish expeditions notwithstanding, and without examining the argument too closely, there does seem to be a certain truth to Barkley's theory of the destructive power of English imperialism. Until the 1920s, when white Anglo-Saxon developers began to overrun the place, the Florida landscape, and even a few of its native cultures, in particular that of the Mikosukkee Indians, remained intact. Now, as Barkley and Kata and I learned on our southward quest, the peninsula would be unrecognizable to Ponce de León. Day after day on our seemingly interminable back-road journey we passed shopping strips, interspersed with housing, interspersed with barren grazing fields. Only in the state parks and national wildlife refuges did we encounter any remnant of the native vegetation of the peninsula.

At one point, about two thirds of the way down the state's east coast, in the town of Hollywood, we came to what must be one of the most irony-laden landscapes in all of America.

Hollywood itself is one of those spots that Kata so abhors—huge clifflike condominiums set in flatlands, interspersed with swamps and upland pines; its once pleasant beaches are now lined with hotels and apartments, and its boardwalk is so famous among French-Canadian holiday-makers that a feature film was once made about it, titled *La Floride.* Just to the north of the city, the state wisely set aside a small green strip and (relatively) pristine beach known as the John Lloyd State

Park. Through a series of strokes of good luck a friend of mine acquired what was essentially the last privately owned green space along the beach and a small, run-down house that had belonged to one of the last fishermen in the area. The house is set in a tangle of sea grapes and wild vegetation, invisible from a small spur road that runs up to the park. It faces the beach, which is no more than forty yards away, and at night veritably shakes when the surf is high. The place is well built though, having withstood several nasty hurricanes in the fifty years of its existence.

Periodically my friend and his wife make journeys to this place through the unholy wasteland of modern Florida. Just before they arrive, they enter one of those vast cool supermarkets and stock up on enough supplies to last for the duration of their stay. Once provisioned, they enter the shaded, leaf-shrouded garden of their property and never go out. The house and grounds provide sanctuary in the truest sense of the word—a place apart, immune from the perversities of the world at large.

Barkley and Kata and I were struck by a period of rain and decided to hole up for a while in the house. We arrived at night in a downpour with slanting sheets of water drenching us, picked our way along the path and the wet leaves to the house, and settled in. That night Barkley found a copy of *The Magic Mountain* and started rereading it, and for the next four days, while Kata and I walked the beaches and the nearby park, explored the squalor of the boardwalk, and swam in the ocean, Barkley read. He read while we made coffee. He read at the table at dinner. He read late into the night. He got up at five-thirty every morning, went for his daily bird constitutional (when he saw more species than Kata and I together in a day of exploration), and then retired to read. Periodically he would emerge to read pronouncements to us.

"'Time in that place is measured by human temperature, by the thermometer, by the course of fever.'

"'Jewish tradition holds that time and place measure the soul.'"

One morning while Kata and I were lounging on the beach in front of the house, he emerged and stood gazing back at the place in its surround of dense vegetation.

"This is a subtropic alternate to the Magic Mountain," he said. "We could stay here forever. We could die here of some tropical disease. 'Space, like time, engenders forgetfulness....'"

Then he went back to his reading.

"He's been getting crazier all winter," Kata said.

Barkley remained so sealed in his reading that he never uncovered the place for what it was. It was very un-Barkley-like. He must have finally been depressed by the real Florida.

Kata seemed to take over his role while we waited for him to finish *The Magic Mountain.* She and I descended to the ninth circle of Hollywood developments. Mostly we would borrow the house bicycles and ride down to the boardwalk to drink a coffee. The French Canadians, who have created a mini-Quebec in this part of Hollywood, have strung their boardwalk with cheap little tourist shops, bars, and small restaurants with wide terraces and outdoor seating. From dawn until late at night, all of Montreal promenades there in a panoply of bizarre costumes, tight spandex, voluminous sweatshirts and T-shirts sprouting advertisements, place names, and dirty jokes, tiny string bikinis, baggy surfers' shorts, and all intermediate stages of dress, as well as undress—Buddhalike male bellies above skinny legs, huge-bosomed Venus-of-Willendorfs, and great ranks of oiled naked bodies basking on the beach in the warm spring sun.

By night the cafés and bars come alive with life. Knowledge of French pays off in that corner of America. Most of the shopkeepers and waiters and almost all the tourists speak French. We actually came to like this place for its contrasts. Whenever we needed to we could retreat to our green sanctuary to join Barkley in his reading campaign. Once ensconced, we could hear the wind in the leaves, the restless surf beyond, the call of birds at dawn and dusk. As long as we did not walk more than fifty paces away from the house, we could exist in the illusion that we were in a more primordial state. But if ever we yearned for a little life of the human sort, we could cycle down to the boardwalk and in five minutes be immersed in the Québecois version of Forty-second Street. What impressed me most was the fact that almost everybody there seemed to be enjoying themselves, drinking and smoking and eating meat with abandon, as if debauchery and a high-fat diet were the best way to hold death at bay. At night the reveling reached fever pitch. Crowds from the cafés and bars spilled out onto the boardwalk itself and the beach was alive with walkers and revelers and lovemakers. Everywhere else we had been in Florida people seemed to be sour and angry and either obsessed with personal destruction or health-conscious. Here they abandoned all caution.

One morning out on the spur road Kata and I were approached by a French-Canadian apparition, a woman in leopard-skin tights with flaming red hair and an orange headband, dripping—as the expression has it—in costume jewelry, mostly glassy crystals. She had two Rottweilers with her and was standing in mid-road staring into the brush.

"Listen to those birds," she said as we approached. "Just listen."

The birds were indeed having a day; catbirds, cardinals, mockingbirds, thrushes, sparrows, finches, and warblers were chattering and squeaking all around us in the remnant native vegetation.

"I adore birds," Madame de Léopard said. "I love all animals, I mean, dogs, cats, gorillas, but birds are special. I believe they are messengers from God."

She had come to Hollywood some years ago for a vacation, she told us, but was so taken with the place she never returned to Canada. Now she owned a shop devoted entirely to natural crystals, a place called "Art by God," which was located on the boardwalk. She insisted that we visit her there, and to get on with our walk, we agreed.

We never did find it, but in the few days we were there I came to understand that this was not just a company of strangers in this place. In many of the cafés and restaurants there were regulars—people actually knew each other. Although garish and noisy, the place had the flavor of a Mediterranean piazza, you could go there alone at night and actually meet people and talk (or try to above the music), and then return to some lonely room and still feel that you are not alone in the world. That, in many ways, is more than you can say for the suburbs of the United States. One goes to work, perhaps to lunch, and then returns to the manor house, the great walled castle, shut in, shut off, from the outside world except through the intervention of electronic devices, television, telephone, and more recently, the networks of computers. What is missing is what the sociologist Ray Oldenburg describes as the "third place," that gathering place away from home and away from work where on a daily basis you can go to meet friends, mingle with strangers, and find lively conversation.

Such places were once the heart and soul of a community's social life. In larger cities they were also intellectual focal points: art movements and revolutions in thought and politics were launched from the cafés of Paris and the coffeehouses of London and Vienna. In some ways, these third places were the defining element of a town or city, they were at once the forge of the character of a given region or place and also its ultimate expression, the focal point in which regional identity would be exhibited. The pub can only work in England, the beer garden in Germany, the taverna in Greece, the coffeehouse in Vienna, the sidewalk café in France, and the coffee bar in Italy.

In the United States the third place was preempted by the myth of the manor house, generally referred to as "home," the isolated, private castle where the lord of the suburban domain held sway over a few square yards of clipped grass, an above-ground swimming pool, a family room with a large-screen television, and also, and perhaps most important, a strong car. That way you can make your escape to the other American institution, the frontier.

"What hope can there be?" Kata asked.

We were sitting in a place called Pierre's with Barkley, having drawn him down from *The Magic Mountain* for an evening on the boardwalk. Kata was getting worked up again about the failure of America.

"You may hate all this, Barkley," she said, "but at least it's a place to be, a center. You go to Centerville, U.S.A., where are you, who are you, where do you even sit down to find the place?"

"The mall," Barkley said.

He wasn't joking. He'll intellectualize anything. "Who says I hate this. You're the one who's suppose to hate trashed environments, that's your job. Mine is to live the life of our epoch. So is yours, actually, you have no choice."

"Who says?" Kata asked.

"Thomas Mann. And anyway, I love this place. This is like the wayside chapel where the questing Grail knights are initiated into the mysteries. It's another wasteland. We have to stop here to get to the paradise of the Everglades."

"He's been reading too much again," I said.

"No, I mean it," he said. "This is good. But tomorrow let's go to the Everglades. Let's charter a canoe and take it up the west coast into the interior and get lost in that green mangrove tangle of channels. Let's sweat and get bitten by mosquitoes, and spend the night on Indian shell middens and listen to the alligators roaring."

"That sounds good," Kata said. "What about your Magic Mountain?"

"I've finished."

MILE EIGHT

Location pertains to feeling; feeling profoundly pertains to place; place in history partakes of feeling, as feeling about history partakes of place.

—Eudora Welty, "Place in Fiction"

Castle Rock 

From the Indian mounds the land rises in a series of small sharp valleys and hills. We pack up our lunch and push on through waist-high blueberry, tall oaks, and white pines until we reach the top of the ridge. We walk south for a while and then turn east and begin to descend towards West Street, which we will have to cross to get over to the Spencer Brook area. Halfway down the hill we pass through an area of woods that has been cut over, probably for firewood, one of the popular home heating fuels in this part of the world. The tract has been selectively cut. There are no dead trees and no older gnarled trees. Brush has been cleared and stacked neatly on the slopes, and a rough track leads down the hill where the freshly cut logs were dragged out. It has all been done with forest conservation techniques in mind, but there is no more mystery here than there would be in sylvicultural planting: the trees are a uniform age, offending species have been removed, and everything is tidy. Sir Henry Sinclair, were he with us now, would be mystified.

For convenience sake, we follow the track down the hill and suddenly come upon what is known in the timber trade as the landing. Here are large stacks of logs piled neatly in rows, waiting to be picked up by the firewood cutter or a truck from a lumber mill. The landing is muddy and rutted and littered with old Styrofoam coffee cups. There is a dirt road beyond the landing leading to West Street, but as we start down, we hear the barking of a number of large dogs and stop to reconsider.

Dogs can be a problem for travelers in the hinterland of American suburbia. On one of our more ambitious pilgrimages, Barkley and Kata and I once bicycled from Cadiz in southern Spain all the way up to South Uist in the Outer Hebrides. We took back roads all the way and passed many dogs en route, but it wasn't until we returned to the United States and took a short ride in the Connecticut suburbs that we were attacked.

Dogs and the concept of private property—the home as castle—have created an unpleasant combination for innocent walkers such as ourselves. Some of the more suspicious natives in this region favor large aggressive beasts and do not always keep them tied up, to the dismay of local deer hunters. Perfectly friendly animals who sleep all day by the hearth and thump their tails when their keepers enter the room revert by night to the Pleistocene and become pack hunters, decimating local herds of deer and also sheep. At best these guardians dutifully bark when you pass within their territory; at worst, they charge and bite.

Barkley and Kata, who as nonbelievers in the private ownership of land are notorious trespassers, have a problem with dogs. They are always being barked at, sometimes get charged, and are occasionally bitten. They both like dogs, but they have developed a healthy respect for these guardians of local territories and know how to read dog signs—which bark is bluff, which is bluff with defense, which is outright aggression. The two in front of us on the road, I would say, are of the latter persuasion, and we have drawn up ranks to assess.

One is a huge brown thing of indeterminate parentage with a black mane, now raised into a mean hackle. The other is a houndlike shorthair with a clipped tail and perhaps a touch of the dreaded Doberman. Both dogs are standing in midroad, sounding off loudly, their tails up, not waving, and when we move forward they back up and prepare to circle for a rear attack.

"Bad signs," says Kata.

Barkley goes off into the woods and returns with a stick.

Years ago in his youth my father lived in China for three years. He too was plagued by dogs on cross-country walks, but he always used to tell me that all of dogdom understands the sign of a raised stick.

Barkley raises his weapon and steps forward. Instead of retreating, Black Mane increases his decibel level and counterattacks.

"False charge," Barkley says, and steps forward again. Black Mane charges a few steps, the shorthair begins its circle.

All this is part of the traditional trials of pilgrims and heroic adventurers. Wild beasts, dogs, and human robbers often waylaid European pilgrims on their way to sacred centers. The route to Santiago de Compostela in particular had treacherous zones outside certain villages. In one account pilgrims described finding bodies of the sick or halt consumed by village dogs.

Dogs also figure in some of the great voyages of mythology. Cerberus, the three-headed, dragon-tailed dog, guarded the gates to the Greek underworld. Aeneas fed him a sop to shut him up in the *Aeneid*. In Dante's *Inferno,* Virgil and Dante placated him by feeding him a handful of earth. Orpheus charmed him with a song when he descended into the underworld to retrieve Eurydice.

Mythical dogs weren't always bad, though. In the *Mahabharata,* the five Panadava brothers, including Arjuna and the queen, Draupadi, began their final pilgrimage to Indra's heaven at the end of their earthly lives. King Yudhishthira brought his dog along. At the Himalayas, all save the king sank slowly into the earth. Yudhishthira pressed on, trailed by the loyal dog. As he staggered, weakened, sorrowful, and blind with weariness, at the very heights of the mountain ranges thunder broke and Indra appeared in a chariot blazing with light and invited Yudhishthira to enter into the kingdom of heaven. "Leave the dog behind," Indra commanded, when the king attempted to mount the chariot with his companion. "Dogs cannot enter heaven." The king dared to argue. "I cannot abandon the dog who has followed me all this way." The two of them were arguing in this manner over the meaning of earthly attachments, when suddenly, in a flash of lighting, amidst choral voices, the dog vanished. In his place stood the shining god Dharma. Indra praised Yudhishthira for his faithfulness and welcomed him into heaven.

In the real world, the traveler Ponce de León owned a vicious hunting dog named Bercerrico whose ferocity was known throughout the Spanish Main. He was such a good Indian fighter that he was allotted the pay, rations, and share of the booty allowed to crossbowmen in the company. The Caribs finally killed him with a poisoned arrow. Ponce de León wept when he died.

"I don't know about these two, Barkley," Kata says. "I don't like them."

She too goes into the woods and comes back with a long stick.

Black Mane and the shorthair increase their volume again, and begin a swirling retreat, coupled with false charges.

I go in the woods and select a stick for myself, an old hickory branch, which with a little trimming makes an excellent club.

"Shall we attempt a frontal assault?" I ask.

"I don't like this shorthaired devil," Barkley says. "He might get behind us."

Kata makes a false charge of her own, with a high-pitched scream, her stick raised.

The dogs become frantic and swirl down the road ahead of us.

Barkley joins the charge now. I walk behind, guarding the rear flank.

We gain ground in this manner, driving the dogs ahead of us, until we come to a clearing near West Street. There, amid a sea of litter, old trucks and old cars, is the house of the woodcutter himself. The dogs take a stand in front of it.

Plato says the dog is the most philosophical of all animals because it has a system of morals. If it knows you, you are good. If it doesn't know you, you are bad. He might have gone on to say that the dog has refined this system to include a sense of place. If you are bad, i.e., unknown, and you are beyond the dog's known territory, you will be warned not to enter, by means of barking and growling. But if you cross a certain invisible line, known to the dog but not necessarily to you, it will attack. Stay outside that line and you will be safe.

The two territorial guardians of the woodcutter's private property, now joined by two smaller dogs of lesser ferocity, are still holding the road. Every time we try to pass they charge out with serious intent. Black Mane, in particular, is judged by my companion dog pyschologists to be a true biter.

"Who was it who tricked Cerberus with a morsel of food?" I want to know.

Kata gets the idea and opens her pack. But we can find nothing that would appeal to a dog save our last two eggs mimosa (which I was actually looking forward to eating later in the afternoon).

"Shall we try?" Kata asks.

She takes one of the eggs out of the container and holds it toward the dogs. Ears cock, the smaller dogs stop barking. Then, with a fine underhand pitch, Kata tosses the carefully prepared egg—stuffed with crushed yolk, Dijon mustard, *fines herbes,* and a dash of dry sherry—to the lowly dogs. Black Mane shies away, thinking it's a rock. The others scramble for it. While they're at it, we slip past, and by the time we reach West Street, we are safely beyond their territory.

This perception of territory, of a base of some sort, seems to be a deeply rooted biological concept. Fish, butterflies, birds, and higher mammals all exhibit varying degrees of territorialism. In the arena, the fighting bull will select in a certain section of the ring an area he seems to prefer above all others, an area known as the bull's *querencia*. If the matador enters his *querencia,* the bull will charge; the rest is bluff.

Querencia is defined variously as a haunt of wild beasts, a home or nest, a favorite and frequent place of refuge, a preferred place, and also as a love of home. In a larger sense, though, the term connotes a deep sense of wellbeing that is associated with a given spot on earth, a sort of personal identity with a place that arises from the fact that the world there is known to you, that its history is your history, that the fruits and flowers, the scents of the earth, the days and nights, and seasonal changes are a part of your personal past. Dissociation from such a place is unhealthy for the spirit, the Spanish say, it leads to unhealthy wanderings and explorations. Columbus, it has been suggested, lacked *querencia*. So did Ponce de León.

The Hopi word *túwanasaapi* conveys the same idea. Though it means literally the center of the universe, it also describes that place which is right for you. The Navajos have an equivalent belief; they model their conical hogans on the sacred mountain, heart of the earth from which they emerged as a people.

It was through the study of birds that the phenomenon of territory was first recognized. Aristotle noticed that birds—eagles, in his case—have territories, as did Pliny. But it was not until 1920 and the publication of *Territory in Bird Life,* written by the retiring British naturalist Eliot Howard, that ornithologists began to take scientific note of the phenomenon.

Eliot Howard's understanding of bird behavior emerged from observation of the birds on his own territory. He was a businessman who worked for Britain's largest manufacturer of steel pipe, near Birmingham, but he lived out in the country, at a place called Clareland. Like many of his nature-loving predecessors, he led a small quiet life. He would rise each morning long before dawn and hike out over the surrounding countryside—

properly attired, one presumes, in sensible shoes and tweeds, with a pair of field glasses hung about his neck. By eight o'clock he would return, change, and sit down to a breakfast, ride his bicycle to the train and arrive at work in Birmingham at the appointed hour—the quintessential small life.

In these stolen hours before work he studied the behavior of the birds around Clareland, the moorhens, wood pigeons, sedge warblers, and tree pipits. Like all good observers he noticed that they did not seem to spread themselves out arbitrarily across the landscape, that certain individuals stayed in certain areas. The fact did not go unnoted in his journals. He was a dreamy man, given to sitting for hours by the fire staring into space. One day he approached his children's nanny and for no apparent reason blurted out the word "territory."

"That's what everything's all about, Nanny. Territory. Territory."

Territory is an area of space that an animal or a group of animals defends as an exclusive preserve. The word "terrioriality" describes the compulsion to defend or possess a space, an inherent drive to hold ground. It is not much of a jump to suggest that the human concept of private property is an animal instinct, a territorial compulsion that worked its way into the genes of human beings a hundred million years ago. If you progress steadily along these lines of thinking, it is not so far-fetched to suggest that humans have an innate desire to defend and even gain territory, and that every tribe, every nation, has a near umbilical bond to a piece of ground that decrees distrust of invaders or anyone who even threatens to cross a border.

On the east side of West Street the land rises again into a series of hills. This again is hard walking, but it is a more pleasant environment. Rocky outcroppings and big glacial boulders are festooned with a covering of leathery rock tripe lichens. Lush green rock polypody and Christmas fern sprout from unlikely crevices and cracks in the rock faces, and sections of the valley floor are covered with wood fern and cinnamon fern. There are mountain laurel, oak, maple, and here and there a remnant juniper left over from the days long past when this land was sheep or cow pasture. At one point the rocks ascend into a veritable castle wall, capped with a squared, crenelated tower. Were this the Levant, a mystic land fought over and fertilized with holy blood, I would say we were passing through the ruins of a Crusader castle. But in fact this is only New England, and the granite strongholds are the work of far more violent historical upheavals; internal earthly fires forged the rock, and the late great glacier wrenched them into the present structure.

It is out of our way again, in fact we are quite off route now, but we climb to the top of the castle rock to get a view of the surrounding woods. There is a flat area at the top, bedded with leaves from last autumn's fall, and we settle here for a few minutes. Below us the land is twisted into a series of sharp, wooded hills and we can see down into the canopy of the surrounding trees on the low ground. In one of the trees there is a vast stick nest, which Barkley inspects with his binoculars.

"Great horned owl, probably," he pronounces. "Then taken over by gray squirrels. Perhaps reoccupied by the owl, maybe a red-tailed hawk. This was the summer nest of the squirrels. Dry leaves. In winter they move into the hollows of trees, where they will raise a second brood, perhaps. They lay low in these hollows when the weather is bad. But most of the time, as we bourgeois bird feeders know all too well, they will emerge to feed. The gray squirrel is a member of the order Rodentia. It feeds on nuts and berries and—"

"Enough, Barkley. We all know about gray squirrels," Kata says.

"He should have been a nature teacher," I say.

"I was a nature teacher once."

"I thought you taught English," I say.

"I did, but I spent most of the time discussing nature. When the Vietnam war came, I objected, conscientiously mind you, and was sent by our national government to test cow's milk by way of repentance."

"It was fate," Kata says. (She and Barkley met while he was living at a dairy farm in the Berkshires. Kata had rented a studio there.)

"Anyway, the school authorities did not like my English teaching techniques."

"All he did was tell his nature and adventure stories."

"The students didn't mind, as I recall. We used to hold classes outdoors, even in winter. There was a woodlot behind the playing fields where there was a fox den. One morning we saw a screech owl in its hole in that woodlot. The whole class got within four or five feet of it. It was the highlight of seventh-grade English for a whole generation, I daresay. But the school officials were not happy about my rambles. I used to quote Wordsworth to the principal to prove I was right: 'One impulse from a vernal wood may teach you more of man, of moral evil and of good, than all the sages can....' That sort of thing. He was not amused."

"The function of the American educational system is to prepare the student for the industrial machine," Kata says. She seems to be quoting.

"Who said that?" I ask.

"Me."

Kata has had her own problems with schools. She makes her living by teaching Native American craft techniques, and while the teachers are her main allies, the funding programs are not always available.

This great castle rock where we are sitting shows no sign of human occupation. That is to say even though there are houses no more than half a mile away, it appears that no child has ventured here. There are no signs of stones modified for fortification, no evidence of encampment. In my time we would have practically lived in a spot like this.

I grew up in a town with cracked slate sidewalks and decaying brownstone estates that had once been a ferryboat suburb of New York City. There was old money in the town and people there tended toward eccentricity; they kept donkeys in ornate Victorian carriage houses and let algae and frogs take over their swimming pools. I think it was there, in the ruins of overgrown backyards, that this quest for understanding *querencia* first began.

Above the town, on the cliffs of the Palisades above the Hudson River, there was a stretch of wildwood, now fortunately preserved as a state park. The woods along the cliff top were dotted with the ruins of former estates which were deserted during the Depression and torn down, leaving a strange, nostalgic landscape of foundations, overgrown garden walks, pergolas, and in one case a ruined tile swimming pool. Every Saturday in warm weather friends of mine and I would ride up the hill to the cliff top and range through the woods, where we would reenact the epic battles of history.

The woods was a wilder version of the town just west of the high cliffs above the river. In the town, among the large 1890s brownstones and Victorian queens, were old carriage houses and horse barns, some deserted and caved in, many with dank cellars where, rumor had it, murder victims lay buried. At the southern end of the town was a great walled estate known as the Baron's, which had a tunnel beneath it and whose inner sanctum we explorers and trespassers never managed to penetrate.

The backyards were large and wooded and worth finding out about, and the exploration of this incongruous environment was the work of our daily life. We would rise in the morning, eat breakfast, and spend the day rambling through the town looking for adventure. In comparison to this realm of possibility, television held no sway. Within this region, a gentle slope of no more than two square miles, we had our special places, our retreats and sanctuaries, known, so we believed, only to us. Some were tree houses, some were well-fashioned caves, but most were located in empty sections of rooms of the abandoned carriage houses and barns. Here we made our own decorations, formed clubs and societies, and prepared for war.

Given the freedom, children will recapitulate history.

The adult community had separated itself into three fairly clear territories: the larger houses on the wooded hill west of the cliff, where the rich and famous lived; below, the warren of smaller houses and streets in the valley near the railroad tracks, where the less wealthy lived; and farther down the hill in the south end of the valley, the Fourth Ward, where black people lived. We children refined these territories even further. We had sections of the hill divided off into known regions, each dominated by a loosely organized group of kids. The hill children lived in fear of a similarly organized band known as the downtown kids, who lived on the lower slopes and in the valley. Periodically the downtown kids would make forays up into the nation of the hill, and sometimes, by prearranged word of mouth, a benign version of a gang war was organized.

These were not anything like the current turf wars of Los Angeles. We armed ourselves with sticks and stones, not automatic weapons, and in fact the battles were more bluff than anything else, a way of restating the known lines of territory. Macaques, baboons, and other primate groups reenact similar territorial battles periodically.

There seemed to be certain seasons of conflict, usually spring and then again in the fall. I remember one spring war in particular, which was rumored around the hill to be the great war. On Saturday afternoon, we were warned, an army of downtown kids would be making a invasion. This was all great adventure. We collected sticks. We laid in a store of heavy timbers and stones in the loft of Candy Templeton's horse barn, and when the appointed hour arrived, two of us climbed into the rickety cupola to survey the hill for the approaching armies.

About three o'clock we saw them, a dark band, moving up Walnut Street, their spears shouldered and pointing skyward. The alarm went up, scouts went out and by shouting taunts and lobbing rocks we drew the enemy into a prepared ambush. Our strategy was to encourage them to collect in front of the barn and, once they were assembled, to pour down a withering fire of sticks and stones. They took the bait and chased our scouts into the barn, but instead of assembling for a siege like a proper invading army, they broke through the doors immediately and scurried through the stalls and lofts with a great racket, dispersing the weaker, more pampered rich children of the hill.

A raging battle stormed through the barn. Some of us held them off at the ladders leading to the lofts with long poles and sticks, hammering at the hands of the climbing soldiers. Some of our warriors moved down a set of rickety stairs from an empty servants' quarter on the south side of the barn and counterattacked. There were sword and stick duels among the stalls, and staffs were wielded in the style of our hero, Robin Hood. On the equivalent of the threshing floor, there was shouting and clamor and brave sorties from the defenders.

But in the end we broke rank and scattered, each running to individual sanctuaries. Some ran to hideouts in sewers and under bridges, some ran home to their mothers, and two of our company, with enemies in hot pursuit, were rescued by my own grandfather, who happened to be visiting. (He was a seasoned street warrior himself. He and his friends in Staunton, Virginia, used to throw stones at a young local professor they called Skinny Wilson. Skinny went on to become President of the

United States and, having learned a thing or two about the nature of territorial loyalties, attempted to found a League of Nations to end all wars.)

My grandfather, recognizing an unfair fight when he saw one, encouraged the two boys to take refuge in our house and dispersed the others.

With our ranks broken, the invading army had but to chase us down individually and beat us up to prove the vulnerability of our supposed territorial boundaries.

My friend Rokey and I lit out for another horse barn on Pitcairn's Hill to take refuge. I still have dreams about that barn, a great cavernous place with dead pigeons and broken roof beams. Three or four bigger boys were in pursuit, but since Rokey and I knew the territory, we were able to lead them astray. We got to our sanctuary and hid in an upstairs room, clambering up a leaning beam to get there, since the stairs were caved in. From the heights we watched the marauding scouts searching for us, and after they had departed, and with evening approaching, we went home to dinner and a warm bath.

Kata has similar memories of her native place, the Tiburon Hills just north of the Bay Bridge. When she was young she used to spend most of her free time ranging the pastures and unpopulated, flower-carpeted slopes that led up to the heights of Mount Tamalpais. Tiburon, now the center of North California culture, was a run-down railroad town when she was growing up. She says the place was vibrant with musical accents, Italian, Yugoslavian and Japanese, and the hills around her house flowed ever upward, an unimaginable wilderness, dense with lupines, poppies, blue-eyed grass, mariposa lilies, sticky monkey, and Indian paintbrush. Here in the hollows and hidden valleys she and her girl club wandered barefoot. They built camps of grass huts and staked out territories, they held meetings and ate ceremonial foods (raisins), and played at being warrior women. It was only later in life, when she was grown and immersed in Native American thought and mysticism, that she came to realize that these same hills, this exact locale, had been the native place of the local Miwok and Castanoan peoples of the Tiburon Peninsula.

There was an old Catholic church at the top of her road, backed by the sweeping hills. She used to hear the bells on Sunday mornings and would look out her window at the lines of parishioners filing up the narrow path to climb the stone steps to attend Mass. One Sunday afternoon on the slopes above the church a giant bird swept out of the empty sky and brushed past her, nearly bowling her over. The creature was larger than any of the turkey vultures that sailed above the grassy peaks. She thought somehow it was a visitation, an unreal bird come down from the mystic peak of Tamalpais, but years later she learned that as late as the 1950s, when she was a child, there were California condors living in the Tiburon Hills.

For years she dreamed of the place.

The coffee houses of Lisbon or Madrid, the cafés and bistros of Paris, the tavernas of Greece, the clubs and pubs of London have their own atmosphere and traditions. They are fascinating places, but they are *places.* You go there with some purpose.

—Georges Mikes, *Coffee Houses of Europe*

Experiences occur in places conducive to them, or they do not occur at all.

—Ray Oldenburg, *The Great Good Place*

Spencer Brook Marsh 

FROM THE CASTLE TOWER Kata and Barkley and I descend and press on to the east. At the bottom of the slopes we come to another small, unnamed stream that runs southeast and appears to flow into Spencer Brook, which, because it is deep and has wide marshes, may prove difficult to cross. Below the castle rock the land flattens slightly, and the vegetation changes from oak and maple to another stand of white pine with very little underbrush. Walking over the ground is easy enough in this section, but the limbs are low and we have to keep ducking under snags and at some points actually have to break through tangles of dead limbs. It's noisy going. Any wild animals we had hoped to surprise will have fled ten minutes before our arrival.

There are many signs of mammals around us. Deer droppings are everywhere, and we find twigs nipped by rabbits, mink droppings by the small stream, and at one point gnawed bark on the upper reaches of one of the hemlock trees, which Kata presumes to be the work of porcupines. We are passing through what must serve as a corridor for the local wildlife, a sort of greenway that the animals use to get from the open space of the Carlisle State Forest to the north and the undeveloped section of land and the Nashoba Brook Wildlife Sanctuary, behind us in the west. This route we are following is a part of a larger corridor of green. If you were a coyote or a wandering moose and were not disrupted by marshes and streams, as we are, and if you picked your route carefully, it would be possible to wander from the greeny hinterlands of southern New Hampshire, south to the square mile of open space of Scratch Flat in Littleton, south-southeast through Acton, down to the Estabrook Woods and on to the Great Meadows National Wildlife Sanctuary, which, give or take a few interruptions, runs all the way down to Sudbury and Framingham. All of this is a distance of twenty-five or thirty miles, and offers one of the last green corridors in the region. Even the major north-south barrier, Route 495, is passable. Several streams run under it, plus a number of small rural roads that are empty of traffic by night. In fact, according to some wildlife biologists, highways are proving less of a barrier than was formerly believed (although I find it hard to accept). Box turtles somehow manage to cross the New Jersey Turnpike, I have read, and recently biologists have found that mountain lions in southern California commonly use culverts beneath superhighways in their wanderings. From time to time animals of the wilder districts do indeed appear in the Concord corridor. Moose have been sighted, coyotes are common in the area, and mink, weasels, otters, and of course deer, are everywhere, and I keep waiting for a peripatetic black bear to wander in from the west. So far the closest they have gotten is Clinton, which is some twenty-five miles to the southwest and is located on another green corridor, which runs west from Concord to Mount Wachusett, an area now under consideration for formal designation as greenway.

This question of where to live when you have a choice used to be a constant subject with Barkley and

Kata before they settled in the Berkshires. The topic comes up often when the three of us are traveling together, and quite naturally, during our recent quest in the anomalous landscape of the American South, we had to discuss the question constantly. Somewhere in Georgia we more or less reached agreement on what we would like in a community. The ideal place we agreed upon was very unlike those cities and towns listed in the "Best Places to Live in America" surveys you see periodically.

We would settle outside a village in a small stone house with a flag terrace and half-wild gardens. From the terrace you have a view of distant hills. The land to the west is unpeopled and wild and rises into sharp, unscaled peaks. To the east, within view from the terrace, is the village, the essence of the place. There is no traffic in this area, in fact we would be happy if there were no cars. You walk to town. In Thoreau's words, you saunter there, poking along as you go, looking at things, listening to birds. The town is small but intelligent. People read books there, and they sit in the cafés and talk about things, and furthermore, they are there all day and late into the night so that whenever you want some company, you have but to saunter along the thicket-lined track into town and find them. And whenever you want the abiding peace of nature, you can walk back to your cottage. If you want wilderness, you walk west to the mountains. Sometimes friends from the village wander out to your house for dinner and you discuss things late into the night, and sometimes they fall asleep on the couches. You find them sprawled there in the morning.

There is a train station in the village and within half an hour or so you can be in a city with a river running through it. The city is a glorified version of the village. The river is clean and swimmable. But the streets are littered with peanut shells and orange peels, and the markets smell of old vegetables and wine. People get drunk and sing late into the night. There are crowded cafés, a few theaters and music halls, cinemas, and a lot of holidays when the people dress up and promenade. There are festivals in spring and autumn, during the harvests, and periodically there are parades when old men wear uniforms and march with canes. The wars are over, though, and we are living in time of peace.

Such places exist. We have been there, but for some reason, we have always moved on. Maybe in the end we are too American to settle.

Years ago I worked in a restaurant called the Rose Café on the north coast of Corsica. The café was actually a small *auberge* with a decent restaurant and a few dusty bedchambers above the dining room and café terrace. It was set on a small island connected to the village by a long windy causeway and was utterly unassuming, a two-story building with French dormers, a wide stone terrace, a pillared verandah, a dining room with a cool bar in the back, and a long narrow kitchen behind the bar. In back of the main building there was a steep cliff that dropped into one of the rocky, surging coves. At the cliff edge there was a one-room stone cottage, where I lived. The house was downright Thoreauvian in simplicity. I had a narrow bed, a table, a candle, and a few wooden pegs on one wall to hang clothes.

It was a good place. You could lose yourself there, you could forget that you ever had a past or a future and simply fall into that idyllic, dreamy state the locals called *la dolce fa' niente,* and within a few weeks I became a sort of adjunct to the place and stayed on longer than I had intended. I washed dishes and cleaned fish, peeled vegetables, helped with the table when the restaurant was crowded, and learned, in passing, the mysteries of the patron's sauces. Other than that I was free. I read books, I went for walks, and at night I eavesdropped on the local gossip. Mostly I stared into space and waited for something to happen. For hours, for days, weeks, finally for months, I simply gazed out across the harbor to the green slopes of the hills and the high, jagged peaks beyond. I rarely left the little island. The Hopi would say I had found my *túwanasaapi.*

The community of people who gathered at the Rose Café that summer came from all walks of life and from all parts of Europe. Some stayed on for only a few days, some remained for most of the season, but almost all, in some small way, left their mark with the small contingent of regulars from the town who, every evening after dinner, would stroll out the causeway to the café to sit on the terrace, talk, and play cards. Separated as we were from the village and the main island, we got to know each other in those long evenings. There really was nothing to do but talk.

I don't know why I left that place, except that I was young. I went back to Paris, where I had been a student, and lived in a room in the little warren of streets behind the Pantheon. I had many friends by that time in Paris and was a part of a floating community of international students and French theater people who would meet every day in the Café Saint-Placide. This was the quintessential third place that was so well described in Roy

Oldenburg's book *The Great Good Place;* you could count on seeing friends there every day, and there were familiars and acquaintances, and there were also strangers. Some of the regulars were hard-working students who earned their board by working as maids and au pairs, some were associated with a nearby theater school, and some were street musicians who slept under the bridges at night.

We regulars would show up at the café in midafternoon, discuss politics, sex, and philosophy, smoke a thousand cigarettes, drink too many *demis,* eat ham sandwiches, and then go to a cheap restaurant for dinner, where we would discuss politics, sex, and philosophy. After that we would go back to the Café Saint-Placide, to continue the discussion until the waiter, Gilbert, closed up. Often we would take a last cognac with him after the place was dark, and sometimes we would walk over to Les Halles, arm in arm, singing sad songs. We thought we were reinventing the world, but of course it had all been played before. Anyway, we had a center, a place to go, and if, after a few days, you didn't show up, people would wonder and talk and make up stories to explain your absence. The place was a place; you went there with purpose.

Just beyond the stream we come to another small rural road called South Street. On the other side of the road the brook opens into a marsh, and so we walk north to get to some high ground we can see from the road. Just opposite the wooded area we see children sitting at a table in front of a house, selling apples. There are two of them, a boy of about twelve with neatly combed short hair and his sister, a younger child of about eight. They have set up little boxes of apples, some jugs of cider, and also individual apples on a card table covered with a white sheet. A sign beside the table announces APPLE FOR SALE.

Kata insists on buying a few.

"How's business?" she asks.

"A little slow, actually," the boy answers.

"Did you grow these?" Kata asks.

"My father grew some," the girl says, "and we bought the cider at the mill—"

"He didn't exactly grow them," the boy interrupts. "The tree grew them. He just picked them."

"I picked them too," the girl says.

"We'll take four," Barkley says and begins fishing for money.

A gray minivan drives by slowly and the boy calls out, "Apples for sale." The van slows but moves on.

"That's what they all do," he says. "They just look."

"Well, in the old days you would have had a lot of people like us, just walking by," Kata says. "They would have bought. Cars don't buy things."

"Cars can't buy, actually," the boy says.

"Are you, by any chance, intending to enter the law, by way of a career?" Barkley asks.

"Never mind, Barkley," Kata says. "This is a good thing to do, sell apples on Columbus Day when everybody is out. I'm surprised you haven't done a spanking business here. Except that there're no walkers. Do you ever go over here to the marshes?"

"Yes, yes," the girl says excitedly. "Our dog, Wolf, he got lost over there and fell in the brook and was covered with mud and my mother wouldn't let him in the house and so we had to wash him with a hose and he didn't like it either."

"He is not accustomed to baths," the boy says.

Another car goes by and the boy calls out again.

"No luck," Kata says. "I'll take four more apples, though."

The girl takes a little cardboard toy box from under the table, opens it, and passes it over to her brother, who begins carefully counting out Kata and Barkley's change. On a homemade ledger sheet the girl carefully notes the amounts received, change made, and profits. When the exchange is over, the boy inspects her accounting work.

"How much have you made?" Barkley asks.

"Quite a bit," the girl says. "Three dollars and forty cents all told."

"Actually, that is not a great amount, Jessie," the boy says carefully.

Another car, another lost customer.

"Maybe we should move on," Barkley says. "Maybe we're bad for business."

"It's possible," the boy says.

I can't imagine what he means, unless he is suspicious of the blue jay feathers in Kata's hair.

"Where are you going, though?" the girl asks when she sees us head for the woods.

"On our way to Concord," Barkley calls.

The boy watches us walk north toward the woods and then stands up.

"That's not the way," he calls. "That way you just go nowhere. You have to follow this road south for one and one half miles and then you'll come to West Street. You follow that and you'll see Pope Road on the right, but don't take it, actually, just go straight, you'll get to Lowell Road, and that will take you down to Concord, over the river. The well-known North Bridge will be over the hill to your left, which is east at that point in time."

Kata comes back to the road and looks southward in the direction he's indicating.

"Our idea is to walk to Concord through woods, without ever touching roads," she says.

"That's an impossibility," our guide says. "Roads are everywhere and you will not be able to cross that big brook over there."

"The one Wolf fell in," the girl says.

"Just what I fear," I say. "Do you know a way across?"

"No. It is not possible to cross without getting wet. You will have to get wet."

"Maybe there's a log," Kata says.

"Perhaps," says our guide.

"Well, we'll just give it a try," Kata says.

Just as we turn to leave an old dachshund comes out from the drive, voices one perfunctory bark, and then stands at the edge of the road staring at us, wagging his tail.

"That's not, by any chance, Wolf?" Kata asks.

"Yes," the girl says. "He fell in the brook."

"His actual name is Wolfgang," the boy says. "His nickname is Wolf."

The contingent of Colonel John Robinson's Westford Minutemen would have been marching to Concord on the Lowell Road by this point in their journey, not far from where we are. We have been interweaving with their route all day. Since they sought out main roads, whereas we are trying to avoid them, they would have been well ahead of us by now. Northwest of us on the other side of Spencer Brook, the Carlisle contingent would also have been heading to the North Bridge. Runners had brought word of the approach of the British to the Carlisle Minutemen before dawn, and by six in the morning they were assembled and out on the common. They started about seven and moved south, toward the place the locals later called Estabrook Country, where there was a lane that led down through the woods and fields, past the farms of William Kibbe and George Estabrook, to the bridge.

Marching in the Westford group that morning were men whose surnames now appear in the local phone book or have been forever enshrined as place names, streets, or town buildings: names such as Joshua Parker, Amaziah Hildreth, James Fletcher, Ephraim Cummings, Silas Proctor, Joseph and John Prescott.

Westford, Concord, and the surrounding Middlesex towns had been preparing for this moment since early 1775. The year before, the Provisional Congress had sent out an order to have munitions and arms stockpiled for the gathering storm, and after the so-called Powder Alarm in Cambridge, when a local stockpile of munitions was raided by the British, Concord was selected as the principal arsenal for the colonial militia. For months, in secret, buried beneath hay and straw and piles of manure in oxcarts, weapons had been spirited into the town. By early spring the tools of war—muskets, shot, cartridge paper, cannon, cannonballs, matches, tents, hatchets, flour, beef, butter, and rum—had been stockpiled in local houses. In town, firearms, gun carriages, and cartouche boxes were being manufactured at an increased rate; guards were posted at the town bridges; and a system of alarms was organized to spread news and the whereabouts of the British regulars. The British knew about all this. They had organized a system of spies and informers and knew what was stored where. One of these men had informed General Thomas Gage, commander-in-chief of the British Army in North America and the governor of the Bay Colony, about the stores at Concord and the mood of the village. By April 15, 1775, General Gage began preparing a company of some eight hundred Regulars for an expedition to Concord to destroy the stored munitions.

Information flowed in both directions, though. Concordians knew what was coming, and the weeks and days before the British expedition were filled with a "dread Suspense," as Ralph Waldo Emerson's grandfather, William, phrased it. But long before the fateful march, tension had been building, farmers carried their muskets to the fields, news and rumor skittered through the towns like dry leaves, and companies of militiamen, organized according to age, class, and war experience, practiced maneuvers and drills on local muster fields. Twice a week on the town common in Concord the minutemen would assemble. On those days the motley crew, uniformed in nothing finer than homespun and hunting coats, marched to and fro across the green, executing the shouted commands of their officers as the fifes shrilled and the drums sounded the paces.

Bliss, indeed, it may have been in that new dawn to be alive—the fiery patriots, the hopeful musters, the old men, and the women, and the children gathering at the common to watch the drills, and the fine young sons with serious expressions, and the seasoned veterans of the French and Indian wars who had stormed the breach at Louisburg in Nova Scotia, and the clipping turns, and the stamping and snorting of the volunteer horse companies, and the fifes sounding out the popular tune, "The White Cockade." You would think there was such democracy and unity among them that such a people could never be broken by the indifferent British regulars, with their paid soldiers plucked from the warrens of London and Dublin and commanded by decadent aristocratic officers who had purchased, rather than earned, their rank. This, the colonials must have thought, would be a quick engagement. It turned out to be the longest war in American history, save Vietnam.

Besides, the colonials were not as unified as legend holds. For one thing, the town of Concord was strictly divided and split by church loyalties, by money and class, and of course by race. Some of the townspeople in this nascent democracy owned black slaves who, when the rumors of war came and the community armed itself, were not permitted to bear weapons for fear it might induce thoughts of a rebellion of their own. The territory of Concord was split into various districts—the North Quarter or the South Quarter—and these districts were not necessarily equal. To some degree, where you lived determined your standing within the community.

Furthermore, when the idea of throwing off the British yoke arose there was the question of loyalties.

Not everyone in the colonies was committed to the idea of revolution, and within the tightly knit colonial communities, where everyone was known to everyone else, disloyalty to the cause meant isolation, verbal abuse, and sometimes outright assault. In Littleton, not far from the site where we began our walk, when a local minister publicly announced his loyalty to the Crown, a crowd assembled in front of his house and fired on it, and eventually drove him out of the town. The same thing happened in Concord to the well-known citizen Dr. Joseph Lee, who harbored some doubt about resistance to the crown. He was roused from his bed one night and "tried" for his views by the committee of correspondence, which had been formed in 1774 for the purpose of municipal self-government. The committee condemned him to house arrest and made it clear that if he went beyond his bounds, he would be killed. One night a group of thirty or so soldiers gathered in front of his door and fired a musket ball at the old gentleman. It was an ignominious act against a respected citizen of the town.

Concord was relatively unified, however. Elsewhere, Tories were tarred and feathered, covered with mud, and shot at, and in Boston they had their houses beshitted and fired upon by angry mobs.

Even in time of peace, the tight communities were exclusionary. Towns took care of their own, and if a local family fell on hard times, ill health, or poverty, the local people would contribute to the family's welfare in one way or another. The custom led to the establishment, in the early nineteenth century, of community-supported town poor farms, where the unfortunate would be housed and fed in exchange for work. But if you were not a member of the community and you thought to settle, and you were in dire economic straits, you might risk a "warning out," as the eighteenth-century phrase had it. At worst a warning out meant that potential paupers—those down on their luck, the idle, and the immoderate imbibers—might be physically removed from the community and returned to the town from which they had come. At best such people were allowed to settle, but were continuously humiliated and shunned.

A strong sense of place begets a strong sense of community. In an ideal situation the community pulls together, cooperates, takes care of its people, develops its village pride, its cuisine and accents, tolerates and supports its local eccentrics and characters, and has in some cases an indigenous music, or a literary style—and, above all, a sense of itself.

I once drove across the three cordilleras of Colombia with a friend of mine who came from a large Cali family and had developed a great appreciation for his country after living elsewhere in the world. ("Life is such an adventure in Colombia," he used to say. "Every place else is too *safe*.") The trip can be done in one or two days, but it took us almost a week because we had to stop so often in tiny, nondescript villages. One was known for a certain style of pottery or weaving, one for a certain type of candy or cheese or soup. At one point in our journey, I fell asleep in a cloud forest and woke up in a hideous, tawny desert. We had changed course so as to visit a place called "La Candellaria," a monastery known for its honey. For hours we plowed through a treacherous eerie zone of dry, ruined villages and finally came to a walled monastery by a river, green with orchards and alive with the hum of bees.

I had another friend, a folklorist and penny-whistler, who spent all his free time collecting tunes, stories, and accents in smaller villages in the British Isles. He had an infallible way of finding the last of everything. He would go into a pub and strike up a traditional tune on his tin whistle and within a few minutes he'd draw a crowd that would inform him of the whereabouts of some old tinker who could tell the old tales or sing forgotten melodies.

This diversity of human landscape, the sense of place, is now threatened with extinction because of the intrusions of the global village created by television and computer networks, the success of multinational marketing, the European Economic Community, the General Agreements on Tariffs and Trade, NAFTA, and other realities of late-twentieth-century economics. The tight little islands where the sense of place endures have, or had, their dark side, however—the isolation within the community of those who happened to be different, the custom of shunning, the suspicion of strangers and outright xenophobia, and, finally, war—the small turf wars of the urban ghetto, the vendettas of Corsica and Sicily, the historical wars of the city-states of Greece and Italy, and finally, the grand wars of place, such as the Second World War, which Robert Ardrey, the author of *The Territorial Imperative* and the archdefender of the importance of territory in human behavior, cites as history's most memorable demonstration of the passions of territory.

We have now crossed the high ground just west of the Spencer Brook marshes and pick our way through brambles at the edge of the woods before entering into a zone of oak crisscrossed, like most of the world in these parts, by old stone walls. A chipmunk dashes into a crevice in one of them when it sees us coming and then, when it thinks we are past, pops up out of another crevice and flicks its tail. Stumps and flat stones in the wall are covered with broken acorn shells and split or shredded pine cones. This, for some reason, is one of those years that produces an inordinately successful nut crop, and the red and gray squirrels and the chipmunks are in a feeding frenzy, filling up their larder for the coming onslaught of winter. As we walk towards the open marshes to the east, Barkley hears what he thinks is a winter finch calling above one of the trees and stops to inspect. Finding nothing he sweeps the trees with his glasses and then halts abruptly.

"By God," he says. "Take a look at this little devil."

He hands me the binoculars and tells me where to look. At the top end of a half-dead oak tree, a little gnome with big eyes stares down at us from its hole: an eastern flying squirrel. I hand the field glasses to Kata.

She searches the tree and cannot find it.

It is at such times that Barkley and Kata play out the drama of their strange interrelationship. Although she sees details of color, form, and animal life close at hand, Kata never can find the birds and distant objects Barkley insists on showing her. In these situations he assumes a certain fatherly patience that eventually enrages Kata and causes her to give up. This time, though, the squirrel intervenes.

While she is searching, it emerges from the hole and executes a balletic leap from the height of the tree. We are standing in a small clearing, and the gray, kitelike form free-falls through the open branches just above our heads. When it is a mere twenty yards above us, it alters course slightly and veers off towards the middle trunk of a young oak, flips its tail downward, lands gently on the tree, and scrambles upward to the next branch.

"So delicate and silent," Kata says.

The squirrel bunches up on a limb, looks back at us briefly, and then scrambles up the tree and enters another hole.

Mile Ten

Every continent has its own great spirit of place. Every people is polarised in some particular locality, which is home, the homeland. Different places on the face of the earth have different vital effluence, different vibration, different chemical exhalation, different polarity with different stars: call it what you like. But the spirit of place is a great reality.

—D. H. Lawrence, "The Spirit of Place"

No place is a place until it has had a poet.

—Wallace Stegner, "A Sense of Place"

Thoreau Country 

In 1521, eight years after his first voyage to find the Fountain of Youth, Ponce de León set out again. This time he had two ships, two hundred men, fifty horses, domestic animals, and agricultural tools. Also, priests. He followed his original course though the Mona Passage, coasted along the length of Hispaniola, and then continued northwest through the treacherous waters of the Bahamian archipelago to Grand Bahama, where he crossed the Florida Straits and then sailed down the east coast of the Florida peninsula. He probably stopped to take on water at a place that would later come to be called Cayo Hueso, or Bone Key, because of the number of skeletons of dead Indians killed by conquistadors that had accumulated there. English-speaking latecomers to the place mispronounced the name and ended up calling the island Key West. From Cayo Hueso, old Juan and his small fleet pushed northward again, once more threading through the broken lands of the west coast of Florida. Ahead of him lay a land of freshwater springs that welled up out of the green earth and ran to the sea in crystal rivers. One of them would restore life.

Key West evolved into one of the most successful communities in the East. In the early part of the nineteenth century, when John James Audubon stopped there during his quest for American birds, it was one of the richest towns in the United States. For the next hundred years it suffered various ups and downs economically, but up until the mid-1960s, when tourists began arriving there in droves, brought by package tours, Key West was one of the few places in Florida—for that matter in the United States—that had a clear identity. Everyone showed up there at one point or another. New York intellectuals, Cubans, blacks, Conchs (the long-settled white families), homosexuals, sailors, fishermen, writers, and motorcycle gang members all drifted down to the place, as if they were attempting to escape the United States of America, and had banked up in Key West, where the continent ends.

One of the centers of the people's community there was a Cuban sandwich shop on the corner of Duval and Petronia streets called Zacaria's. In earlier times, Zacaria's was the typical third place, the spot where, daily, people would collect to gossip and discuss world events, baseball, and music. Sometimes musicians would gather there, and on certain evenings, Saturdays in particular, the crowd stayed late into the night, singing, listening to the music, and sometimes jeering the performers.

Years ago, when he was more footloose than he is now, Barkley lived in a strange dwelling called the House of Hope on Petronia Street. In those times, the island was divided more or less into various districts: the tourists had Duval Street, gays settled in neighborhoods just to the north, the Cubans had another section, the Conchs had theirs. In the late 1960s, when Barkley first went there, black people were relegated to Petronia Street. It is not now, nor was it at the time, perhaps the safest place for an innocent New England white boy, but for that very reason it attracted Barkley, and he found lodging at the House of Hope, a locally run free hotel where the Petronia Street regulars could spend the night for little or no money whenever they needed to. He liked the place and the people, and what's more, they liked him; he ended up spending the winter there, helping the owner with the few chores that were required to keep the place habitable.

On our way south to the Everglades from Hollywood, Barkley got it in his head that he would like to go back to Key West to look up some of his old friends at the House of Hope, and since I had a friend who had a restaurant there and had offered us a place to stay, we decided to take a side tour. It was two hundred miles out of our way, but by then we were already off-route by some five hundred miles.

Barkley was looking for a friend of his named Skin who had been a fellow worker at the House of Hope. But Key West had changed drastically, and the House of Hope was nothing more than a dilapidated ruin, its doors ajar, with broken windows and weeds in the tiny front yard. We spent the afternoon walking up and down Petronia Street among the domino players, asking after Skin and Jimbo and some of the other former residents of the House of Hope, but no one had even heard of the place, let alone any of its residents. Finally we found a decrepit old black man with a slouch hat and a cane and asked him. His answers were right out of *Heart of Darkness.* "Skin?" he pronounced, when Barkley mentioned his name. "Skin? He dead."

He laughed privately and shook his head in remembrance.

"How about Jimbo?"

"Jimbo. He dead too."

"William?"

"William dead now. They all dead."

We turned around and drove to the Everglades and spent the next night camped as close to Florida Bay as we could get. In the middle of the night, in the over-arching stillness, I heard dolphins breaching. The next morning, to cleanse ourselves, we rented a canoe and followed the narrow, uncharted channels up into the Fox Lakes and the empty expense of saw grass. The grass was hot and smelled of salt, wood storks wheeled in gyres above the islands of hardwoods, and lines of ibis drifted over the horizon. Ahead, to the north, the slow river of grass stretched a hundred miles to the first line of decadence, the thin ribbon of the Tamiami Trail, which runs from Miami across the Everglades to the Gulf Coast. Somewhere to the west, across the saw grass, in the Gulf of Mexico, in 1521 the two caravels of Ponce de León and company beat slowly northward. In this emptiness one would not be surprised to see his billowing crowd of sail and the red banners of the House of León.

The marshes along Spencer Brook are relatively dry, so we cross directly through them to get to the edge of the brook. Closer to the stream, we come to areas of standing water in the marsh and again have to leap from hummock to hummock of tussock sedge to get to a small stand of black alder at the stream edge. At this point Spencer

Brook turns out to be about twenty feet wide and shallow with a dark muddy bottom and periodic drifts of grasses, twigs, and bright red and yellow maple leaves floating on the black waters. There are signs of muskrat and otter along the banks, and in the clear sky above the marshes, wood ducks whistle and sail in little groups of two and three.

"No crossing here," I say. "Let's go upstream."

There is a small island of high ground north of us, and still high-stepping on the grassy mounds we gain the dry land and take out our untrustworthy, ill-dated map.

Somewhere around here we have probably crossed an arbitrary line, drawn and redrawn in the centuries since 1635, that defines the town of Concord. More likely we are now in Carlisle, which was carved out of the original territory of Concord in the eighteenth century. Nothing surprising here: this whole region was carved out of the original Concord settlement, and, metaphorically at least, so was the whole United States—students of the history of law credit the town of Concord with creating the first democratically written constitution, among other historical events. It's also the first place in America with enough belief in a government of law to hang a white man for the murder of an Indian.

Had we been hiking on roads, we would have been informed from time to time as to which town we were in, but the untracked woods in these parts is more or less terra incognita; no one ever goes there but deer and grouse hunters. Despite the lack of signs, we are now coming into a territory loosely referred to by everyone from real estate sales people to environmentalists as "Thoreau Country." We are probably within the range of

Henry Thoreau's daily walking excursions, some of which took him as far afield as Littleton.

Most well-known naturalists come to be associated with a certain region or place: John Muir with the Sierras, John Burroughs with the Hudson River Valley, Aldo Leopold with Sand County, and of course Henry Thoreau with Concord; the two—man and place—are almost inseparable.

It was Thoreau's custom, especially while he was living at Walden, to rise in the morning, take stock and write for a while, and then in the afternoons set out through woods and fields to see what he could see. He undertook these excursions in all weathers, in all seasons, and he delved into habitats that others studiously avoided: swamps, thickets, forests, rivers, marshes, and the unfashionable and outlying regions of Walden Woods and the Estabrook Woods. He would commonly log ten to twenty miles in a day or, by contrast, would spend hours high on a brushy hill collecting blueberries with children, or immersed in swamp waters watching frogs. He walked the woods, he walked the Concord railroad tracks, he walked the streets of the town, the gardens, the fields, the farms and wood lots, streams, and hills. There was not a corner of the community he did not know intimately, thanks in part to his personal quest to, as he says, travel a good deal in Concord and in part to the fact that as a sideline he gained his livelihood by working as a surveyor. There never was a more passionate pilgrim, a deeper explorer of the wilderness of the nearby than Henry Thoreau. Like a devout Hebrew, his whole life was pilgrimage.

We see a point on the map where the brook narrows considerably and leave our island and hike down to it, still carefully selecting our route to keep dry. By siting distant high spots and then hiking to them, leaping from hummock to hummock, and braving dense alder thickets, we again reach a section of solid earth beside the black stream. Here a thin, slippery elm tree has fallen across the brook, providing a potential bridge. I venture out onto it, but when I'm halfway over it begins to sway and then, ominously, sink with my weight. "Turn," they shout from the bank, and precariously I execute a spin and start back. The curving trunk rises up behind me slowly as I exit, as if to indicate that the wages of sin lead but to the hell of branch water. John Bunyan's earnest pilgrim, Christian, often came up against similar barriers in *Pilgrim's Progress.*

We are forced to reconnoiter again.

Henry Thoreau was a dedicated naturalist, especially in his later years, and he went about his explorations in a scientific manner. He was a faithful note keeper and journal writer; scholars who have delved into the two million or so words that make up his journals find evidence there of major nonfiction works in the making. One of them may have been a treatise on American Indians. Another was a study of the dispersal of seeds. In 1993 the manuscript for this unpublished work was edited and printed as *Faith in a Seed.* The ideas describe the process of plant succession and anticipate current theories of ecology by nearly one hundred years. Thoreau integrated Darwin's theory on natural selection with his observations and applied it to plants to come up with what is now recognized as an accurate description of the way in which plants distribute themselves in a given habitat. This kind of accurate scientific observation is one side of Henry Thoreau's mind. But there was more than scientific method to his observations. Thoreau saw in the natural world around Concord a grand metaphor for spiritual themes, in his case a manifestation of transcendental realities in common landscapes.

The revelation and record of this grand metaphor produced what is, as far as I know, the first book devoted entirely to the exploration of the idea of place. There had been a few precedents. In the mid-1700s, the Englishman Gilbert White explored the natural world of his own town in Kent in his book of letters, published as *The Natural History of Selborne.* Romantic poets, Wordsworth, in particular, commonly sought inspiration and metaphor in natural phenomena. But none of these, nor any previous works of fiction or nonfiction, explored so intentionally and so thoroughly all aspects of the concept of place as *Walden* and Thoreau's journals.

Having failed to find a crossing we move north again, away from our destination. We could just walk south to the Lowell Road to the Middlesex School and Bateman's Pond and from there hike into the Estabrook Woods and pick up one of the south-bound trails—but that is far too rational a choice, the point here being to get to Concord in a seventeenth-century landscape, not by the shortest or most convenient route possible.

And so we slog northward again, looking for a crossing.

Margaret Mead once suggested that there are certain places on earth where, for no apparent reason, things happen. Books are written in or even about these places, they inspire great works of art, writers, philosophers, and their followers collect there, she says. Clearly one such place is Concord; another is the landscape around Taos, New Mexico, which drew both D. H. Lawrence and Georgia O'Keefe and many others; and another, somewhat larger, magnetic area is Provence, especially the region around Mount Ventoux in the Vaucluse.

One has but to go down the list of creative people who have either lived in or been inspired by this section of the South of France to understand the effect that geography can have on creativity. By the thirteenth century, the Provençal troubadours were an integral part of the culture of the region and had developed a recognized literary style. The seminal Italian poet Francesco Petrarch was influenced by the region; he loved Provence (although he hated Avignon) and was one of the first people to climb Mount Ventoux. In the fourteenth century, Dante passed through the region known as the Valley of Hell in the Alpilles on his way to Les Baux and used the collapsed rock and cataclysmic landscape as a model for his landscape of hell (although there is a curious, singular mountain in the Apennines that locals there claim is the real model). By the seventeenth century, another folkloric regional literature developed, and by the eighteenth century traveling English, seeking sunlight, discovered the region.

People were always flocking to Provence to heal themselves of another powerful place—Paris. Madame de Sévigné, Daudet, and Dumas were associated with Provence, as of course was Van Gogh, who thought Provence the happiest place on earth and insisted that all the cicadas there sang in ancient Greek. It was the light of Provence that drew Van Gogh. It also drew Cézanne, Matisse, Dufy, Braque, Derain, Chagall, Picasso, Nicholas de Staël, Gauguin, and Renoir, to name but a few. The power of the place endured into the twentieth century, after the region was "discovered," and popularized as a tourist area. Serious artists, filmmakers, and writers, such as Jean Giono, Marcel Pagnol, Jacques Prévert, Roger Vadim, Marcel Carné, Colette, and their numerous sycophants and followers continued to flourish there, as did American expatriates, exiled English kings, White Russians, and of course twentieth-century versions of the traditional Provençal smugglers, pirates, and gamblers. By the 1950s nearly half of all of French culture seems somehow to be associated with Provence.

For a brief period during the mid-nineteenth century, Concord had a similar concentration of creative minds, albeit smaller. But for so young a place—Provence had a thousand years to gather its artists—the list is impressive. Drawn by the charismatic Ralph Waldo Emerson, who returned to his ancestral territory to live in 1834, other writers and thinkers began to visit or even settle in Concord, so that by 1840, this small satellite of Cambridge and

Boston had become the American center of intellectual activity. Nathaniel Hawthorne settled in the Old Manse in Concord in 1842 with his wife, Sophia, and immediately was fascinated by the town: there was something about the river and the views from the Manse, the light, and the glimmering shadows around the grounds that gave the site an aspect beyond the material world. Until late in his life he referred to Concord as Eden. Bronson Alcott, originally from Connecticut, began living in his Orchard House in the east end of town in 1848. His daughter, Louisa May, used the place as setting for her immensely popular *Little Women,* published in 1868. The poet Ellery Channing lived in Concord. The feminist educator and writer Margaret Fuller visited often, and the presence of Emerson inspired a host of lesser lights, including the sometime schoolteacher, pencil maker, surveyor, and handyman named Henry Thoreau, whose oeuvre was all but unread until after his death in 1862.

Kata has it in mind for some reason that if we hike north on dry ground, we may be able to cross Spencer Brook upstream. There is nothing on the Geological Survey map to indicate that this is true, but, as she points out, the map is useless anyway, and so, even though it is the wrong direction, again we hike north, away from Concord. Once out of the marsh we gain high ground and break out of the thickets into an old field of thin poverty grass dotted with juniper bushes, and an old wagon road running through it. We follow this for a while and at one point come to a strange nineteenth-century barn, isolated and unconnected to any nearby dwelling. It is a weathered, center-entrance structure, its doors half opened, with a weedy ramp leading into the threshing floor—just the type of place that attracts Kata and Barkley. Inside, illuminated by the dusty, raking light, is an old horse-drawn hay wagon with a broken axle, its bed filled, inexplicably, with old kitchen chairs. The walls are hung with moldy harnesses and reins, and piled against a side wall are a stack of fruit baskets and a few wooden crates. Above, on either side of the central floor, sagging haylofts spill ancient, useless hay.

"Good place to take shelter from the storm," Barkley says.

"What storm?" I ask.

"You know, *the* storm," he says. "Late in the day, the light fading, sky darkens, and the cloudburst breaks. We take shelter. Great pounding thunder, flashes of brutal white light, no relief in sight, so we decide to spend the night—"

"Yeah, right," Kata says. "So we go up in the hayloft and settle in—"

"And then," Barkley says, "in one of the flashes, we see *it.*"

"Right, and the resident barn owl breaks from its perch and flutters at the window."

"Let's go," I say. "Before this gets serious."

"Well, it's a good place."

"Evocative," says Kata.

"Aren't they all?" says Barkley.

Once the three of us did take shelter from a storm in late afternoon in an old stone barn in the Swaledale in Yorkshire. We were not the first to do so: nearly one hundred years' worth of notations had been inscribed on the walls by through hikers on the Pennine Way. No ghosts though, and not even a resident barn owl.

We walk north, and then, guided by Kata's intuition, turn east again for the brook.

Although Concord managed to concentrate writers and thinkers, the phenomenon of a region or place fostering creativity in its sojourners is not unique; there are other places on earth that have such a draw. There are perhaps geographical reasons for this. The sea, the presence of rivers, high peaks, and the like have been the traditonal inspirations for poets and painters. But they don't necessarily live in these places, they only visit. A more common attraction may be the presence of low, rolling hills—the hills of the Vienna Woods, for example. Beethoven and Schubert both roamed the hills and forests of the region. The Pastoral Symphony was inspired by the place; so, in all likelihood, was Schubert's piano quintet, "The Trout." The three-quarter-time dance known as the waltz, which in its time was as popular as some modern-day rock fad, was the descendent of the country dances of local peasants of the forests and vineyards around Vienna. Gustav Mahler, Hugo Wolf, and Arnold Schönberg knew the region well, as did Freud—he hated the Vienna Woods, but some of his most significant theories were evolved from dreams that were played out there. Even that tortured, quintessentially twentieth-century man, Franz Kafka, was familiar with the woods. His few fleeting hours of happiness may have occurred there during walks with his lover Milena. And then, of course, there is Mayerling. The popular, liberal-minded, dashing young Prince Rudolph selected the Hapsburg hunting lodge at Mayerling in the Vienna Woods to carry out the double suicide with his lover, thus fixing the place forever in the romantic folklore of the West.

There is a similar hilly region in the south of China, not far from Canton, known as Gui Lin. It is a bizarre landscape of humped karst mountains, rushing streams, forests, and mists. The region drew poets and landscape painters from all over China and appears repeatedly in the classic Taoist paintings in which a small human figure beside his hut is dwarfed by a receding vista of vast mountains giving way to an empty, misty sky.

And then there is the so-called Tuscan miracle. The green, town-topped hills, the quality of light, the olive-studded slopes, the vineyards, and forests of Tuscany somehow combined to create in the residents of the region an artistic sensibility. There is nothing sublime in the region, nothing of the dramatic Alps, and yet it was here, in this landscape, that perspective was discovered; it was here, in the city of Florence, that Giotto, Masaccio, Paolo Uccello, Leonardo, Michelangelo, Brunelleschi, and Donatello worked their trade. It was also here that the paragons of Italian literature came to florescence: Petrarch, Boccacio, Machiavelli, and, of course, the master Florentine, Dante Alighieri, who maintained such a love/hate relationship with his native city after he was banished from the place that he peopled Inferno with his enemies by way of revenge.

The powerful aura that emanated from the hills of Tuscany also influenced embryonic nineteenth-century American authors in search of a style. James Fenimore Cooper, Nathaniel Hawthorne, Emerson, William Cullen

Bryant, Longfellow, William Dean Howells, and Mark Twain all spent time in Florence, most of them staying in the quarter around the wide Piazza Santa Maria Novella. It is perhaps no accident that after they returned to America, having discovered the power of place in Europe, many of these writers became associated with a given region.

But even without the Florentine inspiration, the influence of place on writers has been particularly strong here in America. Historically, the economic climate in the United States seems to have inexorably led to the destruction of places and their pasts, and so it is not surprising that a nostalgic regional literature would develop in reaction. By the 1820s in America a loose association of writers consisting of James Fenimore Cooper, William Cullen Bryant, and Washington Irving had acquired the label the Knickerbocker Group because of the strong identification of these writers with the landscape of rural New York. The term was lifted from a mythical Dutch historian of New York named Diedrich Knickerbocker who was an invention of Washington Irving.

Mark Twain wrote *Life on the Mississippi,* his nonfiction recollections of a passing way of life on the river, when he was in his fifties, and then, having discovered the place as a place, wrote the classic American regional novel *Huckleberry Finn.* Edith Wharton used the wide lawns and terraces of the estates on the eastern bank of the Hudson River Valley as a setting for her fiction; Willa Cather, the windy northern plains. Faulkner used as a setting for his work the diverse human landscape of his hometown of Oxford, Mississippi; Lafcadio Hearn and George Washington Cable, the creole street culture of

New Orleans; Thomas Wolfe, Asheville, North Carolina; Robinson Jeffers, Carmel, California; Sherwood Anderson took as his raw material the small lives of the insular Ohio communities whose only visible means of escape was the railroad track. Even in the anonymous landscapes of the late twentieth century novelists draw inspiration from specific environments. William Kennedy, for example, hated Albany, New York, left for Puerto Rico, and there, at a tropical remove, came to realize that the only place on earth that had any meaning for him was the warren of North Albany, where he grew up.

But of all these places of the imagination, New England remains the most powerful. I once prepared a list of writers influenced by the Massachusetts landscape and had to start crossing off names because there was not enough room on the page to print the complete list. No other region on the entire North and South American continent has produced so many writers in so small a territory. And it all seems to have started in Concord. As Van Wyck Brooks's 1936 book, *The Flowering of New England,* makes clear, it was not just a few literary giants; there were hundreds of popular writers in the region, most of them unread now, in our time. Marcia Moss, doyenne of the Concord Public Library, estimates that some four hundred published writers have lived or worked in Concord in its short history.

By the 1850s, tourists of a literary or historical bent were coming to Concord to visit Emerson, see the site of the battle at the bridge, and view the scenes set down in Hawthorne's *Mosses from an Old Manse.* After Thoreau's death and Emerson's eulogies, and encouraged by his friends and allies, the pilgrims began coming also to Walden Pond to walk the landscape created by Henry. By 1860 Hawthorne was living at the Wayside. Walden, which was once the unfashionable, unvisited woodlot south of the town, was populated by pilgrims and pleasure seekers, many of whom brought stones from afar to lay at a cairn at the site of Henry's humble cabin. Bronson Alcott's *Concord Days,* published in 1872, helped further shape the image of the town, and by the end of the decade his articles and traveling lectures drew even more people to his Concord School of Philosophy. A year later Ellery Channing published a biography of Thoreau, and by 1873, a popular guidebook by Samuel Adams Drake stated that no other town in New England was so well known to the world in general, mainly because of its concentration of writers. In 1875, the centennial celebrations of the Concord fight focused even greater national attention on the place. Every spring, "like robins," as one of the Alcotts suggested, the pilgrims and tourists would show up from parts west to tour the town or attend Alcott's School of Philosophy. The area had become, in the words of one journalist, "hallowed ground"; its very hills, rivers, and trees, were sacred.

Concord was not, it should be said, just another summer colony. There is a thin line between tourism and pilgrimage: the former is a secular version of the sacred journey, the latter is an act of religious devotion. For the pilgrim, the very journey may be part of the ritual. The pilgrim comes to be uplifted, and the Concord visitors of the late nineteenth century were there to be enlightened, not entertained. There was no landscape to draw them; as Samuel Adams Drake pointed out, Concord had no "scenic features sufficiently marked to arrest the tourist." The people did not come for ocean breezes and lakeside boating (although there was some of the latter by 1860 at Walden Pond). These pilgrims were here for an image, they were attracted by the concept of a "native place." Concord, in effect, was an idea, a creation of its literati and publicists, an attempt to establish in the vast amorphous, and by 1850 thoroughly explored continent some sort of center to which the American public could cling, a site that they could identify as uniquely American. The pilgrims wanted an equivalent of the Eternal City or the City of Light, an Athens they could identify as the cradle of democracy, a place to go to prove they were a people. Emerson gave them the image in "Concord Hymn," which was inscribed on Daniel Chester French's statue of the minuteman at the North Bridge in 1875 for the centennial celebrations. Emerson saw the bridge as covenant, the American compact, a symbol of the rite of passage of the new nation: the rude bridge, the embattled farmers, the shot heard round the world. After 1875 we had a native place, and when, in the film *Northwest Passage,* the starving, battle-weary member of Rogers' Rangers goes mad, and announces that he is going home to Concord, the American populace, no matter how unread, would understand the metaphor.

Mile Eleven

The Muse, disgusted at an Age and Clime,
Barren of every glorious Theme,
In Distant Lands now waits a better Time....

—Bishop Berkeley, "On the Prospect of Planting Arts and Learning in America," 1752

In this age, when a meagre utilitarianism seems ready to absorb every feeling and sentiment, and what is called improvement, in its march, makes us fear that the bright and tender flowers of the imagination will be crushed beneath its iron tramp, it would be well to cultivate the oasis that yet remains to us....

—Thomas Cole, "Lecture on American Scenery," 1841

Estabrook 

Kata believes that the Hopi people with whom she lived were prescient. She says they would often see things before they happened and would sometimes warn her of an impending danger. Since she always took their advice, she never knew whether they were actually right, but later, when she returned to the East, she came to think that she herself had developed some of their clairvoyance. Now that we are once again thwarted by the waters of Spencer Brook, she tells us that if we just keep walking northward we will come to a stone bridge across the brook. She can see it, she says, and has learned from the Hopi to trust her inner vision. So we turn east and enter once again the marshy floodplains that surround the brook. After two or three hundred yards, traipsing again from hummock to hummock on the mounds of tussock sedge, we come to the spot she indicated and find an old stone bridge. The stream beneath it is a mere ditch, though, and Barkley doubts that we are crossing the main branch of the Spencer Brook. We walk eastward for a few hundred yards and see that he is right. Once again, we come up against the deeper, wider, waters of the main stem.

"Upstream or downstream, Kata?" Barkley asks. He has always had trouble with clairvoyancy, especially Kata's.

She stares at the opposite bank. "Maybe we should wade?"

The waters are shallow at this spot but one test of the bottom indicates an endless abyss of mud. One local told me some years back that there is quicksand along the banks of Spencer Brook and that cattle were often mired and consumed by it. The black oozing mud beneath the clear waters looks suspiciously bottomless, so we turn and hike south again along the banks, thrashing through thickets of alder and sedge. Then, because the stream begins to widen once more, we turn and retrace our steps and forge northward.

By now we are so far north we decide to march up to a narrowed place shown on the map just below a pond. I was informed, by the same man who told me the quicksand stories, that there was an old mill at the north end of the brook around this spot, and I presume that we will be able to get over the mill dam; barring that we can continue on and—in failure—cross the brook in Carlisle on the Old Lowell Road.

More thickets, more sedge, more oozing waters. We are getting tired. We have spent a good half an hour or more leaping onto hummocks to stay dry. We have slipped and soaked our feet. Barkley's new boots are sogging with every step; we have been scratched and torn by alders, and the sun has that dry autumnal slant that burns your cheeks and dries your forehead to a fever. Kata's cheeks are flushed and her hair has come loose again. A pine stub stabbed my thigh a couple of miles back and we would love an afternoon nap, but it's too wet to lie down anywhere.

Never mind, all this is the work of pilgrimage. For the true pilgrim the journey is as much a part of the pilgrimage as the arrival at the sacred center: a means of absolution, an expiation and purification. Pilgrims do not travel for the sake of comfort.

Before the miracle of air flight, the trek to Mecca across the empty quarter and through the deserts of what is now Saudi Arabia—one of the hottest sections of the earth—was the forge where the true pilgrim was fashioned. For the "marathon monks" of the Tendai Buddhist tradition, the pilgrimage consists of a one-hundred-mile *run* around the sacred Mount Heii in Japan. Pilgrims traveling to the sacred Narmanda River must walk for two years through India. The one-hundred-mile walk to the Black Madonna at Jasna Góra in Poland involves blistered and bloodied feet, cold, rainy campsites, and cramped legs. Old women in poor shoes make the journey. People sing all the way to keep their spirits up. The point of a traditional pilgrimage is hardship, endurance, a cleansing. What's a few hours of suffering on a fine autumn day in New England? And anyway, at the end of the journey we intend to end up at a restaurant. True pilgrims never eat, they feed on spirit, and mortify the body.

And so we forge on.

About two hundred yards below the Lowell Road, Spencer Brook narrows to a fast-running stream, and a little farther on we hear the rush of falling water and come to the mill dam. Since it appears to be part of someone's private lawn and is carefully landscaped, we politely circle around it, slip through a narrow strip of wild vegetation on the south side of the mill pond, climb onto the road, and hike up a hill and then turn southward in the woods.

We are now in one of the richest cultural landscapes of the entire Thoreau Country, the notorious, embattled Estabrook Country.

On the afternoon of October 20, 1857, Henry Thoreau set out on one of his many explorations of what was then sometimes termed the Esterbrook or Esterbrooks Country or Woods. It was a brisk Novemberlike day with little whitecaps forming on the Concord River and a strong wind out of the northwest. Late in the afternoon, on his way up the old Carlisle Road, Thoreau met an old man named Brooks Clark coming down the path in front of him. The old man kept a woodlot of yellow birch trees in the north end of the Estabrook, where he would cut his winter supply of heating fuel. He was walking barefoot, nearly doubled over at the waist, and was dressed in a frock coat so tattered it hung in strips around his knees. His pants were equally shredded, and his knobby shins and bare feet protruded into the cold air. Thoreau stopped to chat, as he always would do with local travelers. Clark was carrying his shoes, and stuffed inside, even up into the toes, were small wild apples he had collected. In one shoe there was a dead robin; he had found it in the woods with a broken wing and killed it for his supper.

Brooks Clark was over eighty at the time, and had, as Thoreau says, "a feeble hold on life," yet he was cheery and talkative, and was especially happy to have found something in the woods to carry home for his winter store. Needless to say, Henry Thoreau discovered a moral lesson in this child of nature. He vouchsafed Clark's wild apples were sweeter than those grown by the good husbandmen of Concord; better his robin than their turkey, Thoreau wrote in his journal. He found Clark's cheeriness worth "a thousand of the church's sacraments...."

This incident took place on the old Carlisle Road, an abandoned road that runs through the heart of the Estabrook Woods. Our intention is to hike eastward to this main north-south trail, and then turn south for an easy walk down to a wooded area east of the Estabrook Road and from there to the Minuteman National Park land to cross the river on the North Bridge.

The land on the western slope of Corly Pate Hill, where we are now standing, consists of highbush blueberry, white pine, and oak—a light, sun-filtered woods with open grassy patches. We traipse along in what for us is easy walking, now that we are on dry land, and about a half a mile into the woods we sit down on one of the stone walls for a taste of wine and a drink of water from Barkley's thermos.

Even in Thoreau's time the Estabrook was deserted. Most of the farms that were located in the area had been abandoned for over fifty years by the mid-nineteenth century, and this section of woods, old pastures, and orchards was considered by Thoreau as one of the large wild tracts of Concord that constituted, in his view at least, the very heart and soul of the town. By his time, the cleared pastures of the Estabrook were growing up to blueberry and birch and wild apples, and in the lowland swamps—some of which never had been cut—ancient trees of the deep forest such as hemlock still grew.

Estabrook is still undeveloped, an empty quarter of

Concord consisting of some five to ten square miles (depending on who makes the definition) of deep woods of hemlock, oak, white pine, and hickory, interspersed with ponds, old pastures, and ridges and well watered with running brooks. Some of the sections of the Estabrook forest are characterized by old trees, some is relatively young second growth that grew up in the blueberry barrens that are still remembered by members of some of the old families who live in the area. At the core of this tract is a section of 675 acres, which is leased by Harvard University as a research area for students of the natural sciences. Another section to the east is owned by the town of Concord, and the rest is in private ownership, including a section on the west side belonging to the Middlesex School. The school has caused a great stir among the conservationists of Concord by threatening to cut out a section of their holdings to develop as playing fields.

The town of Concord and Harvard University maintain a number of trails in the area, much favored on weekends by people from the surrounding towns. On the east side of the woods, just below Punkatasset Hill, where on that April morning in 1775 the minutemen collected before marching on the bridge, there is a wooded pond. Always a favorite local picnic site, for a few years in the late 1970s this was a popular gathering spot for local nudists, some of whom came there on their lunch breaks from computer companies as far away as Route 128 to the east. Shortly after their appearance at the pond, Concord residents began to notice that local contractors, electricians, and telephone company workers were developing a keen interest in nature. On lunch hours during summer, their trucks would line the entrance to the trail that led to the pond, and the journeymen themselves would descend the trail to observe, so it was rumored, the beauties of nature.

Among the nudists were a number of young homosexuals, and although they were discreet, their presence, along with the increased popularity of the site among the nudists, the rise in numbers of voyeuristic workers, and the resulting parking problem, caused a reaction among local family groups who also walked in the woods around the pond. As a result, yet another anomaly was created. During the warmer months, armed uniformed police in shiny boots and polished leather belts, looking very out of place among the trees and naked people, would also come to the area to disperse the transgressors.

Nowadays the crowd has thinned, although from time to time one may still see a lone, naked *sadhu* meditating on a sunny rock by the pond's edge.

The woods to the north have also attracted a number of locals who, temporarily down on their luck, have found shelter and solace in some of the tarpaper shacks and foundations that were constructed over the past decades by those in similar circumstances. For a while, I heard, there was a virtual community living in the north end of the tract, some of whom would wander down into the streets of Concord by day to sit on the walls around Monument Street and watch the parade of passing tourists. Now, with the changing economic climate and the gentrification of the community, these folk have been extirpated, as have the old dogs that used to visit the town, and the sidewalk grocery stores, and the greasy-spoon diners serving high-fat dinners at noonday.

Thoreau said in his journals that every town should

have a park or forest of five hundred or a thousand acres in common possession forever, preserved for recreation or instruction, where never so much as a stick of firewood should be cut. The Estabrook Woods fits this description perfectly. In fact, Henry specifically cites the Estabrook Country, as he called it, as a perfect spot for such an endeavor. He did not, could not, have foreseen that much of the land around the town in which he lived would be so stripped of its native vegetation, so overdeveloped, that this and the other last remaining wild tract, Walden Woods, would be overcrowded with visitors on weekends and holidays.

Since today is one of the various national holidays that attracts these city-bound pleasure seekers to the woods and fields, we find the Estabrook in heavy use. Our first encounter is a near-death experience with a mountain bike.

We have now gained one of the through trails leading to the old Carlisle Road. The walking is easy here, and we soon fall into a dreamy late-afternoon saunter, Kata in the lead, strolling languorously along with her thumbs hooked into her right pack strap, Barkley at the rear, still upright and alert for the calls of birds. A veery has been chipping in the lowlands to our right, and Barkley claims to have heard more winter finches, and a flock of yellow-rumped warblers, also tree swallows, robins, the cry of blue jays, crows in the distance, and, from time to time, the honk of Canada geese ahead of us in one of the ponds.

Suddenly in the midst of this peaceful interlude we hear a shuffle, followed by a rich Barkleyan epithet, as a silent mountain biker sweeps past at high speed and strikes Kata's pack. The rider is professionally attired as if for racing in a shiny spandex body suit with many garish stripes and a bright polyurethane helmet. He pedals off down the trail without apology.

No sooner have we recovered from this first intrusion, than another biker slashes past, and then another.

"Sorry," this last one calls out, breathlessly. "Got to keep up."

Now we are subjected to another one of Barkley's many fantasies.

"Someday," he announces grandly, "I am going to charge one of those things, jam a stick in its spokes and then attack the bike—not the rider, mind you—attack the bike with my knife, slash the tires, beat the frame with a rock, and run off into the woods howling and laughing maniacally. I want mainly to see the headlines: 'Crazed Hiker Attacks Mountain Bike, Leaves Rider Stranded.'"

We expected people in the Estabrook, but this has been a rude introduction. We recover, however, and with Barkley acting now as rear guard, we march on towards the bridge.

Thoreau's monumental, sustained essay on the sense of place, known as "Walden" was published in 1854—and was promptly forgotten. He was one of those authors who was better known after his death. Although Thoreau was notoriously independent, his work was not conceived alone. He was influenced by Emerson and by his readings in Eastern literature and philosophy, and he was very much a part of the Concord scene and the first glimmerings of the environmental movement in the American mind. The phenomenon began, not coincidentally, just before the mid-nineteenth century, just as the Western frontier was opening and the nation was becoming industrialized, and it had two epicenters. One of these, Concord, was literary, and the other, the Hudson River Valley, was pictorial.

In 1825 the young English painter Thomas Cole made a sketching voyage up the Hudson River. Even by 1825 the valley was tamed and pastoral, hardly wilderness, but just to the west in the Catskill Mountains, Cole found a wild as yet unpainted landscape, with high craggy peaks, precipitous valleys and falls, and a dark forest that ranged ever westward to the Great Plains. Cole had studied painting in Philadelphia and had spent three years in Europe before returning to America, and like many of the Romantic painters of his time, he was attracted to the noble, ancient scenery of the classical Italian landscape, with its inspiring ruins and its tamed, pastoral vistas. But after his return to America, Cole discovered a land that was new to art.

His metaphorical landscapes of the Hudson River Valley found willing viewers in a swiftly urbanizing nation, and his work drew a number of fellow artists to the region. Asher B. Durand, John Frederick Kensett, Jasper Cropsey, all of whom had also been trained in the academies of Europe, began following Cole to the Catskills' valleys and mountain slopes with their sketch pads. Although they painted outdoor scenes, these artists were not plein air painters. They would make sketching trips into the wilds and then return to their studios to construct immense, narrative canvasses depicting, sometimes with exaggeration, the pastoral landscapes they had witnessed. Although they never thought of themselves as a group, the painters were given a name, the Hudson River School. They spawned in midcentury another school of American landscape painters led by Frederick Edwin Church, the German-trained artist Albert Bierstadt, and Thomas Moran, all of whom concentrated on the awe-inspiring wilderness of the American West.

In 1836, having found his *querencia,* Thomas Cole settled in the place, on the west bank of the Hudson in the town of Catskill. His disciple Frederick Church had by the 1860s ranged far beyond the valley and was best known for his vast canvasses of sublime, terrifying scenes: the Andes, icebergs, and visionary portraits of the American West. By the early 1870s Church was one of the most popular and also one of the richest painters in America, and when he decided to build his dream house, he returned to the wellspring of his inspiration. He selected a height on the opposite bank of the Hudson River from the home of his mentor and there constructed a great neo-Persian manor with sweeping greenswards and grand vistas down to the river. This was not simply a dwelling place for Church and his family; the site was carefully chosen for what the Chinese geomancers, had they been consulted, would term *feng shui,* a sparsely treed height above water where the air and the wind were propitious. Furthermore, Church designed aspects of his own life and travels into the house, layers of meaning and metaphor. Students of the design of the place, which he named Olana, see it as one of his greatest works of art.

These early landscape painters, Church especially, were moved by what was then known as *terribilità:* the sublime, even frightening, views of a world untamed by human industry. Much of the painting of this genre was metaphorical. In some ways, having a landscape of this group on your walls was not unlike ownership of a sacred icon, a symbolic representation of the deity. Thomas Cole, in particular, saw deep religious lessons in the edifying landscapes fashioned by the Creator. Another group of metaphorical painters, later termed the Luminists, also used outdoor scenes, especially the play of light, to convey the transcendental philosophies that were so much a part of the Concord phenomenon.

Painters such as Cole, Bierstadt, and Church had provided the public with an alternative view of the virgin land. Whereas prior to the mid-nineteenth century the public saw only economic uses of land, by the second half of the century, thanks to painters of the American landscape, people started to see spiritual and esthetic values in wilderness. In a nation obsessed with material progress, it was a big change, and as a result, the American public began to develop an embryonic environmental consciousness. By the late 1800s, when it was proposed that wild tracts such as Yellowstone should become national parks, the public was already familiar with these places. It is doubtful that the preservation of a wilderness park would even have been considered if the painters hadn't been there first.

By the latter part of the nineteenth century in America, the great age of vast canvasses depicting immense wilderness scenes gave way to offshoot schools of landscape painting such as the Luminists and, later, the American Impressionist movement, with its smaller, more intimate paintings. By the turn of the century, a closer, more humane, school of painters began to evolve. Among these was a group not far removed from Concord known as the Boston Ten, who concentrated on local scenes, families, and even cloistered, Vermeer-like interior landscapes.

Yet another impressionist group confined most of its subject matter to a small, few square miles in Old Lyme, Connecticut, along the banks of a tributary to the Connecticut River. The painter Childe Hassam created one of the classic American impressionist paintings in this region, *Bridge at Old Lyme.* Hassam was associated with the painters who collected in Old Lyme around the turn of the century at a dilapidated, late Georgian inn owned by the doyenne of the group, Miss Florence Griswold. The artists who gathered there developed a uniquely American style of painting that blended the brooding landscapes of the Barbizon-school, light, and an intensely colored impressionism. They were largely influenced by the character of their surroundings: the river, the marshes, and the green hillside woods dotted with wild laurel and granite outcroppings.

There is nothing unusual about seeing landscape as metaphor and then creating a work of visual art to represent the idea. The Greek temples were a manifestation of the god in a given locale. Chinese Taoist painters depicted in their landscapes the interplay of the elemental forces of creation: vast mountain slopes, a gnarled tree, a small hut, a smaller human figure, subsumed by receding ranges of mountains, waterfalls, twisted pines, all leading up into a vast, amorphous, empty, sky, the symbolic nothingness of the Tao. A group of fifteenth- and sixteenth-century painters in the Danube basin, known as the Danube school, employed local landscapes to express the interplay of nature with human affairs. Flemish painters of the same period began to solve the problem of depicting spatial reality by painting foreground, middle distance, and far distance, and by the seventeenth century, landscape painters such as Claude Lorrain and Gaspard Poussin were creating paintings in which countryside itself was the central theme.

Except for one obscure work that is attributed to the Luminist FitzHugh Lane, Concord does not appear in any of the American schools of landscape painting—not physically at least. Metaphorically the town permeates the whole genre. Thoreau, Thomas Cole, Frederick Church, the Luminists FitzHugh Lane and Martin Johnson Heade, and most especially Ralph Waldo Emerson, all tended to express spirit, the illusive presence of God, or the transcendental Oversoul, as manifested in physical nature. Never mind that they used different media; the phenomenon of place as expression of ultimate meaning infuses all their works. The art critic Barbara Novak has said that Martin Johnson Heade painted what Emerson thought. Conversely, Thoreau's *A Week on the Concord and Merrimack Rivers* could be read as a Luminist painting, or for that matter a Taoist landscape: the narrator ascends the Concord and Merrimack rivers, climbs through the foothills, ascends Mt. Washington (the sacred mountain of the local Indians), and then, at the very summit, describes nothing, a caesura, an American version of the Taoist concept of *wu,* the emptiness where all things are contained, where the transcendental Oversoul exists. There is even a blank space at that point in the narrative in early editions of the book.

On the north slopes of Corly Pate Hill (a modern misspelling of Curly Pate) we climb to the crest and then descend to the Carlisle Road. On the way down the trail to the road we come upon a benign group of dog walkers—two women in jeans and flannel shirts, surrounded by a herd of dogs. In the vanguard are a group of shaggy mutts of border collie blood, followed by a slower-moving golden retriever, a small, energetic Jack Russell terrier (a popular breed here in horse-pasturing Concord), and then the women, who have three dogs in tow, two on leashes, one on a long rope. Two of these are wandering terriers, and the one on the rope is a white West Highland, who proves to be very friendly. Barkley and Kata have to do a lot of petting and admiring before we are able to move on. Fifty yards farther down the trail we see another member of the pack, a poor, limping black dog with a red collar. He passes us with a worried glance and trots onward.

A troupe of young men and women in mountain parkas and hiking boots pass by, followed by a huffing jogger, and then another dog walker, a woman dressed in a winter coat in spite of the warm afternoon. A few minutes later we come upon an older couple in tweeds. The gentleman touches his Tyrolean hat as we pass, but does not look up. All these sojourners aver that this is a splendid autumn day, a fine day to be in the woods, a beautiful afternoon, a nice place. By this time we look worn and ragged, our shoes are soaked, Kata's sweater is festooned with burrs and seeds, and Barkley's parka is torn, but no one questions our presence in this place. Estabrook Country permits, even attracts, eccentrics.

It also promotes mystery. Early one misty autumn morning a friend of mine was walking with his female companion in this section of the Estabrook when they heard distant thunder and felt the ground shake. They turned just as a wild-eyed white horse galloped out of the mist and swept past, its jaws slavering. Hard on its track, giving chase, was a large white dog. The two raced by, the hoofbeats diminished, and silence returned.

Mile Twelve

There is something sad about the look of the land. One never sees an acre gained from the forest; around the pasturelands there is often a belt where the wood marks its gain upon the cultivated tract.

—Nathaniel Shaler, 1869 (quoted in *American Space,* by John Brinkerhoff Jackson)

Mink Pond 

JUST BEFORE WE REACH THE OLD CARLISLE ROAD, in the untracked woods we see a well-attired woman carefully picking her way through the brambles and blueberries. She is wearing a gray wool skirt, a blue boiled-wool jacket, gloves, and not very sensible shoes. Her dark hair, which is tinged with gray, is full and swept back and is tied in a loose bun in a turn-of-the-century style, and she is wearing expensive silver earrings. She seems to be approaching us for help, and so we wait for her to make her way out of the thickets.

"A mystic?" Barkley asks under his breath. "Our resident genius loci?"

It turns out she is lost.

"I've left my car parked at the Middlesex School and can't seem to find my way back," she tells us, apologetically.

Since we are headed to the trail that will take her over to the school, we invite her to join us and we hike along chatting for a while. She is from out of town, she explains, but used to live in Concord years ago and is now revisiting some of her favorite haunts. In spite of the fact that she has been lost for an hour, she is pleasant and relaxed and seems happy to be with us. She introduces herself simply as Jane, we introduce ourselves, and we walk on, looking at plants and chatting about the beauty of nature, the lovely day, and other mindless pleasantries.

Jane stoops periodically to pick up a colorful leaf—she already has a fine collection in her pocketbook, I notice. She has round, very blue eyes, and when she looks at you she seems to be focusing on something beyond. It occurs to me that she's slightly mad; she remains vague about where she has come from exactly, and seems to have no direction.

Two hundred and twenty years ago on April 19 a different group walked this same road. Militia and minutemen from Carlisle had been aroused by the same organized system of express alarms that had alerted the entire region and by early dawn the town had sent down a small contingent of twelve men to join the gathering companies at Punkatasset Hill. The contingent—historians are reluctant to term so small a group a company—left the town common just before dawn, marched east, and then turned south on the Carlisle–Concord Road, which followed the high ground and ridges through the swamps and dense woods of the Estabrook Country. They were presumably the same sort of raggle-taggle band of men between sixteen and forty-five years of age, dressed all in homespun, and carrying by way of weaponry a variety of hunting pieces and muskets as those who were at this same hour converging on Concord from as far away as New Hampshire and Connecticut.

Paul Revere, who never made it to Concord that morning, was not operating alone in that bright dawning. He and other members of the various committees of correspondence had organized an elaborate system of spreading news throughout the countryside, and although Revere was captured by a group of scouting British sentries who were out that night for the express purpose of stopping such alarms from spreading, other riders had picked up the news and were at work.

Concord itself was warned by one of its native sons. Twenty-three-year-old Doctor Samuel Prescott had been courting a maid in Lexington that night, and was on his way home when he met William Dawes and Paul Revere on the road near the Lincoln-Lexington town line and offered to help out. The three were subsequently spotted by a British patrol and told to stand, but at one point in the confusion, Prescott, who knew the countryside, spurred his horse across a meadow, leaped a wall, and escaped into the woods. Concord bells began to ring shortly thereafter; riders went out, and word reached Bedford and Acton, Lincoln and Littleton and Carlisle.

Ironically, the route the Carlisle contingent took that morning was more populated in their time than it is today. In 1775 there were at least three or four families living in the general area, including descendants of Thomas Estabrook, for whom the parcel was named. Much of the land in what is now deep woods was probably clear—small home lots or intensively farmed plots around the house, cultivated fields beyond, beyond them pasturelands, and then an area known as the "waste" or wasteland, which was basically woods or woodlots, where fire wood was cut and where hogs were turned out to forage in late summer and autumn. The flatlands in the plot, of which there are not many, would have been cleared first, then the hillsides, and finally, if ever, the deeper swamps and wetlands.

Concord farms were originally larger holdings of some two or even three hundred acres. But under the system developed in New England, a section of the property would be cut out and given to the oldest son, and then to subsequent male offspring, sometimes leaving no land for the youngest and, in larger families with a lot of boys, small, insignificant parcels for middle siblings. In order to accommodate ambitious farming children, families would purchase or otherwise acquire unsettled land outside the community, the closer lots first, and then farther and farther outlying lands and so on to the western frontier—which in those days might be no more than twenty-five miles west. In this manner the whole of New England was settled.

By the time of the Revolution, Concord lands had been intensively farmed for nearly one hundred and fifty years. The soils in the area were depleted, the town population was burgeoning, and the distribution of land was no small problem. In Concord as elsewhere in New England, the great western exodus was beginning. In fact, in some communities it was over fifty years old.

Departures are not necessarily well documented, but there is good evidence that 1816 might have broken the back of Estabrook. In 1815 the great volcano Tambora in Indonesia blew its top, one of the largest eruptions in recorded history. The great sulfuric clouds of debris and ash rose some four or five miles into the stratosphere, effectively blocking the warming rays of the sun and causing what came to be known in colder climates as "the year without a summer." The results were felt worldwide, but here in New England the effects were especially troublesome, since the soils were wearing out and the hardscrabble hilltop farms and marginal areas such as Estabrook were already hard-pressed. After the Revolution, there were rumors of better, more fertile land beyond the western frontier in what is today New York State. In 1825 the Erie Canal connecting the Hudson River and Lake Erie opened and the trickle of westward migration began to flood.

The Estabrooks, the Kibbes, the Clarks, the Browns, and the other "outlivers," as they were called, who inhabited the poor farms in the tract that would come to be known as Estabrook Woods, were not immune to this pattern of settlement, and one by one, for varying reasons, the families pulled up stakes and went west. Commonly in those times one did not put a house on the market and wait for a buyer. Families simply closed the door for the last time, climbed in the waiting wagon and slowly drew off. In the case of the Estabrooks and the Kibbes, it was only a matter of time before the swallows and the phoebes moved in beneath the eaves. Raccoons broke into the attics, foxes dug into the cellar holes, winter snow sagged and finally broke the back of the roofs, the walls gave way, and slowly, in the damp March winds, the wooden frame, the clapboards and siding, returned to the earth. Beyond the home lots, the pastures grew up to birch and choke cherry, yew and blueberry, and by Thoreau's time Estabrook was a haunted land, the farms deserted, the families departed, and only a wind blowing.

Halfway to the turning for the Middlesex School we come to a track which leads off to the eastern side of the Estabrook Woods, to the area where the pond and Punkatasset Hill are located. I know this region fairly well, and although it is a wide, indirect loop, the trail will eventually bring us back to where we want to exit on the south end of the tract. Since it is a far more interesting route than the Carlisle Road, we agree to take it, and I give directions back to the school to our lost soul, Jane, expecting her to continue on and go home. Instead she asks if we would mind if she joins us.

"Please do," I say, "provided you don't mind *him.*"

Barkley has just heard some errant fall warbler and is standing waist-deep in brush making loud swishing noises to attract it to him.

The trail we take skirts the southwestern slope of Hubbard Hill, dips into swamps, crosses a small brook, and meets up with other trails at the western end of the infamous Hutchins Pond of local nudist fame. The forest consists mostly of oak, with a few white pines growing among them, some of which may have been seedlings in the years when our hero Henry Thoreau walked this land. On a sunny rock just above the low-lying Mink Pond we see a great fat water snake warming itself in the autumn sun. It slithers across our trail as we approach and heads down the slope for the pond. Jane, for all her apparent urbanity, does not flinch; in fact, holding her hands close to her waist, she follows it down the slope a way to get a better look.

"What a beauty," she says.

She collects more leaves, inspects twigs, snaps off a yellow birch and offers it to Kata, who politely expresses surprise at the fact that it has a wintergreen flavor (something she has known for years). Whenever Barkley stops to search for his birds our guest joins him, happily takes his binoculars to look, and listens with seeming interest to his interminable nature lectures. Inspired by his new student, he begins stopping for even more birds—blue jays, titmice, chickadees—species he would normally only glance at briefly. I watch Kata to see if I can spot the green eye of jealousy in this burgeoning nature bond, but if anything she seems to welcome their shared interest. It takes the pressure off.

From time to time we pass fellow travelers. Dog walkers, three polite boys on mountain bikes, another dog walker, who is frustrated because her golden retriever, Elizabeth, will not heel as instructed, two gentlemen with close-cropped hair, neatly attired. Then, just west of the pond, we come upon two couples with packs and binoculars, staring into the trees.

"See any good birds?" Barkley asks, spotting, he believes, kindred spirits.

"Not much," one of men answers. "Had a black-throated blue over by the Hutchins Pond, and the black-polls are still moving." He is a dark-complexioned man, dressed in light clothes and Teva sandals in place of hiking boots.

"Today's their day," Barkley says.

"Had fourteen in one tree," his new compatriot answers.

While the two discuss birds, we mere novitiates stand around and nod. One of the women in the group has long gray hair tied in a braid and has what appears to be a handwoven pack, probably from South America.

"I like your *bolsa,*" Kata says. "Did Indians make it?"

"Yes, I think, I got it in Ecuador, while I was traveling."

"Otavalo market?"

"Yes, yes. You know the place?"

"I looked for baskets there once," Kata says.

One of the men looks familiar to me, and we begin talking about the Estabrook trails, the owners of various properties on the outskirts, the price of land in the area, and, inevitably, the threats to this beautiful parcel, of which the most recent is a development proposed by the nearby Middlesex School. It turns out we have a lot of friends and allies in common, some of whom are engaged in the fight to preserve the woods.

After about ten minutes of conversation, since we are headed in the same general direction eventually, we join forces with this group and, taking a southwesterly trail, hike on towards the bridge. They have been out all day as well, it turns out, having left from a house in Lincoln and, following conservation land, hiked to Walden, where they had lunch, continued on to Concord, and then looped up through the Estabrook. Now they are headed back to Concord.

The man I had recognized is one of those local Concord fixtures, a sometime illustrator and jack of all trades who does small carpentry jobs for rich families in the town. He is blond, bearded, and is wearing much-worn khaki trousers and torn sneakers. His companion, or wife, Randall, the woman with long braided hair, is a local potter at a Concord artists' cooperative and grew up in the area, just outside the Estabrook Woods. The other couple, Diane and Steve, live in Lincoln not far from the Sudbury River and appear to be archdefenders of local environments. Diane, who is a newcomer to the area, once did battle with a local landholder who was filling wetlands along the Sudbury River to park his construction machinery. She got nowhere. The man was a part of the old boy network of builders and developers of Concord and Sudbury. Phone calls were made, arcane bylaws were cited, and the case was dismissed.

"I should have known," she says, after relating the story.

"But never give up," says her friend, Randall.

"Never," says Kata. (She's being polite; she generally leaves political battles to Barkley.)

"'We shall fight them on the beaches...'" says Barkley.

The others glance at him politely and carry on with their stories of crimes against the Estabrook.

Except for the property owned by the town of Concord, and the lots belonging to Harvard, there is no reason this whole grand forest could not be stripped and developed. Little by little over the years, according to our informants, builders have been nibbling away at the edges, and landholders have been cutting wider and wider trails to accommodate their horses. One of these recently cut a vast highway through the forest so he and his horsemen could ride three abreast. "These are essentially scoundrels," the carpenter explains. "They are polite, and rich, and well educated, and support all the good causes, but in the end, they are scoundrels."

Kata and Randall, the potter from Concord, have found a number of common interests, not the least of which is their relationship with Native American people. Randall, it appears, had been fighting against the huge hydroelectric plant in James Bay and had spent time in Canada among the Cree. As part of an international program to develop local support for the Indians' attempts to stop the massive government-sponsored project, Randall had invited a group of them down to Concord to speak on the subject at various programs. (True to tradition, Concord's people—not its town government necessarily—have a way of connecting with national and international issues. Concord was one of the few communities that protested the Gulf War, the town's citizens were involved in exchanges with citizens in the former Soviet Union, and Concord still has a sister city in Nicaragua and another one in France.)

Recently, Randall tells us, she invited some of her Cree friends to come to town to drum up support and appreciation for Concord's own waters, the Sudbury, Assabet, and Concord rivers. The Cree were supposed to stay with her and then meet on a Saturday morning to commence a great armada of canoes that would float down the Sudbury River to the North Bridge. Our friend was a little worried about putting up the Cree at her house because she knew they loved to smoke and she was allergic to cigarettes. (Kata suffers from the same affliction but endures. "After all, Indians invented the habit," she likes to say.)

Our friend met her companions the night before the big event at the Colonial Inn in the center of town. They had a drink and then set out for her house—the Cree packed into their car, slowly following her. At one point she noticed they were no longer behind her. She circled around town for an hour trying to find them and then finally went back to the inn to see if they had returned. There they were having another round of beer. "Old hunter's trick," they told her. "When you lose one of your party, just go back to the last place you were together."

When the Cree finally arrived at our friend's house they studied the place carefully. "Where's the TV?" they wanted to know. "Don't have one," Randall said proudly. There followed a certain amount of incomprehensible discussion in the Cree language, and then the spokesmen for the group announced that they would prefer, if possible, to stay at a motel. There was a big Canadian hockey game on that night that they did not want to miss.

The flotilla was scheduled to leave from a launch site on the Sudbury River at ten o'clock the next morning, and at the appointed hour all the celebrants arrived, except for the Indians, who were supposed to start the event with a ceremonial invocation. Half an hour went by, then an hour, and the Cree still had not arrived. Our friend, who understood their sense of time, assured the assembled that they would come, but that time as measured by minutes and hours was not important to them. Finally, two hours late, the Indians showed up. Unlike the good suburban conservationists, who were properly attired in outfits from the L. L. Bean and REI catalogues, the Cree were dressed in street shoes, polyester shirts, and shiny pants. (Same way they would dress to run their trap lines and hunt moose, our friend explained.) When it came time for the invocation, none of them seemed quite sure what to do. There was another discussion in Cree and French and a certain amount of foot shuffling, and then one of them stubbed out his cigarette on the ground, stood on the bank, raised his arms and said some words about the Great Spirit. Then he said the Lord's Prayer.

"Amen," says Kata.

Our new companions know the interior of the Estabrook better than I and suggest we take yet another side track through the woods over to the site of the Thoreau family sawmill. We leave the trail and weave single file through patches of poison ivy, brambles, and alder thickets beside a small running stream. Our resident lost soul, Jane, enthusiastically plunges through all the mud and squalor in her fine little leather shoes and gray gloves and long skirt, without complaint. She has been listening attentively to all the Indian stories and accounts of environmental destruction, and she follows along, staring at the trees around her with her surprised blue eyes. It's almost as if she's seeing the world for the first time.

We climb a bank and then, still in thick vegetation, come to the edge of a small wooded pond with a couple of large cedar trees at one end. Nearby on the stream we arrive at a small stone wall beside the stream.

"This is the spot," the carpenter says. "John Thoreau had cedar cut and milled around here somewhere."

When Henry was an adult and a regular denizen of this forgotten tract of land, his father's mill was already in ruins. At one point in the accounts of his rambles in the Estabrook Country, he says he walked over to the site where "father's mill used to be located."

In the last decades of the nineteenth century, when Concord was firmly established as an American pilgrimage site, the whole New England countryside had taken on a sad prospect. Visitors from the city would seek out the countryside and return to report in nostalgic newspaper columns of encounters with lonely swains and white-bearded old men leaning on their scythes like warriors on their spears. As early as 1857, Thoreau's journal accounts of the Estabrook country have a bit of this air. He tells of old Brooks Clark, the fallen hero, carrying his quarry home to his old wife. He writes of the overgrown mill dams where his own father cut and milled cedar for his pencil factory. He writes of empty, boulder-strewn fields and the deserted lime quarry and kiln that used to be active in the Estabrook Country but was in ruins in his time. Over on the western side of the lot, Thoreau reports, there was an old farm with a classically eccentric uncle, driven over the edge perhaps by the changing economic climate of Concord. It was his intention, it was said, to dig through to the other side of the earth. Every night he would pick and shovel for as long as one candle lasted. He never made it, and by 1857, when Thoreau explored the area, foxes had occupied his hole. As late as the 1970s locals could still find a remnant of the pit.

None of this was happening in isolation. Concord was at the center of what the historian Bernard Bailyn has termed the greatest population movement in early modern history, the vast transfer of a populace, westward across the Atlantic, and then westward yet again across the American continent to the Pacific. This was an immense migration of animals in a very short period of time, a massive *displacement* (to use a loaded term) of a formerly settled population from one region to another, and then onward to yet another, and all at very great expense of human life and of local ecology. It's no wonder that we have a problem in this country when it comes to settling in and developing a sense of place; our ancestors must have been impatient, restless, unhappy people to undertake so vast a trek.

By 1840 there were more people on the move in the

United States than in any previous period, and by 1880, in New England, the matter of deserted farms was becoming an increasing concern among journalists, reformers, and sentimentalists alike. The plain living, high-thinking people of New England, who were in essence the model for the nation's way of life, its national character, were giving up their heritage and moving west to uncertain existences. The move went against the popular notion of the American family farm with its white picket fences and prancing horses and dogs, its happy families, its rounds of plantings and harvests, and its wholesome, simple pastimes. The settled image was mostly apocryphal in any case, and the custom of pulling up roots, of wearing out farms, was entirely in keeping with the American character. In many ways, we are the culture that invented the habit. Place holds no sway in the American soul; when things get rough, we get out.

And so we who were left behind, or who came later, weave through a forgotten, uncultivated land as we slouch towards Concord. We wade swamps, cross woodlands that were once cultivated fields, stumble over rock walls with deep histories, and old foundations, beds of periwinkles and daffodils, and the remnants of holes where eccentric uncles once attempted to escape this hellish day-to-day existence by digging through to China.

Fix your eyes on the greatness of Athens as you have it before you day by day, fall in love with her, and when you feel her greatness, remember that this greatness was won by men with courage....They gave their bodies to the commonwealth and received, each for his own memory, praise that will never die, and with it the grandest sepulchers, not that in which their mortal bones are laid, but a home.

—Thucydides, *The History of the Peloponnesian War*

They came three thousand miles and died
To keep the past upon its throne.
Unheard, beyond the ocean tide,
Their English mother made her moan.

—James Russell Lowell, inscription on the grave of British soldiers at the North Bridge

Punkatasset 

SINCE PUNKATASSET HILL IS NOW SO CLOSE AT HAND we decide, by way of commemorating events of April 19, to ascend to have a look. Randall knows the region well, having spent the better part of her childhood in the area, and once again bushwhacking through untracked sections, we head southeast to the base of the hill.

When she was a child Randall used to spend a lot of time in the Estabrook Woods on her horse and would swim nearly every summer day in the now forbidden waters of Hutchins Pond. At one point on our way we come across a small stone foundation of what appears to be an outbuilding located in the woods with no relation to anything.

"This was old man Nutter's place," Randall explains. "He was a Latin professor from Harvard and he used to spend his summers here. I think my grandfather actually owned this cabin but gave it to him outright."

Professor Nutter was a crotchety old man who wanted nothing more than to be left alone with his Catullus and his Virgil. Whenever the local children were making too much noise down by the pond he would appear on the slopes on the south bank with a stick and shout at them to pipe down. Sometimes Randall's grandfather would have the Latin teacher over for Sunday dinners, and at these times, when they were out of earshot, the old man would turn to Randall and whisper, "If I had a daughter like you I'd lock her in an attic room and throw away the key."

"I never knew whether he was joking," Randall says. "But I was terrified of him for years. So was my mother."

The Westford company of minutemen, whose track we have been crossing periodically, was lagging far behind the action that morning. But the Carlisle contingent, whose route we have also crossed, was here among the other minutemen and militiamen who gathered at the top of the Punkatasset Hill.

Early in the dawn, after the alarms had spread, about one hundred and fifty men collected on Concord Common waiting for further news from the alarm riders. By this time—early morning—the provincials had gathered at Lexington Green under the command of Captain John Parker, who had fought in the French and Indian wars and had once been a member of Rogers' Rangers—the same company whose fictionalized madman insisted on going home for dinner in Concord in *Northwest Passage.* The group of colonials had gathered at the common some four or five hours earlier and had been waiting in the cool night for the regulars. As they waited, they could hear alarms signaling the approach of the British, the sounds of church bells and guns firing off through the darkness. Some of the men grew weary and went to the nearby Buckman Tavern to lift a glass or two to take the chill off, but by daybreak, one of Parker's scouts rode into town with the news that the regulars were at hand. Parker's drummer sounded the alarm and some seventy men lined up in double file at the west end of the green.

When he was captured earlier in the night by the British regulars, the wily Paul Revere had informed his captors that there were no less than five hundred to a thousand men waiting at Lexington and ready for a fight. Furthermore, all along their route as they came closer to the town, the British had encountered armed provincials on the roads and in the nearby fields, all of them headed for Lexington. One stopped to warn the British that they would never be able to enter the town. Another, according to British accounts, aimed at them and fired. American historians claim it was an alarm shot.

As a result of all this, when the first six companies of British marched into Lexington under the command of Major John Pitcairn, their weapons were primed and loaded and the soldiers were ready for action. No one really wanted a fight. Pitcairn rode up to the standing force and told them, in so many words, to disperse—what he said exactly is still disputed, but he may have called them "damned rebels" or some other inflammatory name. Parker told his men to disband and some of them began to draw off, although they were in no rush about it.

In the dim light both sides overestimated the numbers. The British thought there were at least five hundred colonials on the common, and the militia and minutemen thought there were some thousand to fifteen hundred regulars facing them. Tension was high, and the fatigue of the march, the excitement of the moment, the half-light, the rum had combined to create an explosive situation. Nevertheless, it might all have ended peacefully, except that someone, no one knows who to this day, fired a shot—which did not come from the ranks of soldiers on either side. As soon as the British infantry heard the shot they began to fire without orders. At first there were a few popping cracks from the muskets, and then a great cloud of smoke billowed out in front of the force with the terrible thundering crash of a volley. There was a great deep-throated shouting of "Huzza-huzza" from the regulars, the traditional British war cry. A few Americans fired back, horses broke and scattered. The spectators who had also gathered at the common ran for their lives; the lead balls whistled through the air; the smoke began to collect and obscure the field; and in the midst of the confusion, the frantic Pitcairn did his best to gain control. Finally, he ordered a drumroll and regained command, just as the body of British regulars who had been following behind the vanguard of Pitcairn's company under the command of Colonel Francis Smith marched into town.

The British reassembled on the common and, on command fired a victory salute into the air, a heavy volley of some eight hundred muskets. Then they gave three cheers and marched on for Concord.

Behind, on the green, eight colonials lay dead.

Punkatasset Hill is lightly wooded now with oak and pine with an understory of low bush blueberry and catbrier. At the top, someone had cleared a few trees, and there are a couple of campfire sites, perhaps the remnants of past reenactments of the fight at the bridge that are played out in Concord each spring. To the east from this height, we can see small planes rising and settling at Hanscom Airfield, another source of the twentieth-century environmental skirmishes that now characterize the Concord climate. Residents have been fighting a running battle to keep the airport from expanding.

The new members of our company have a good supply of food with them, and inasmuch as our lost soul has not eaten in a while, they offer her some cheese and crackers. Barkley still has a little Sancerre in the bottle and he offers her a taste, but she refuses. She also refuses some delicious-looking homebaked chocolate chip cookies, an apparent specialty of Randall. Jane does accept an apple from Kata, and then shyly explains that she suffers from a disease, hypoglycemia, which prohibits the consumption of sweets. Sympathetically, Kata tells her of a similar condition which she believes she once had. Then they began discussing yeast infections; fainting spells; bursitis; vague, unspecified attacks of nausea; wrist problems; and, seemingly, all the slings and arrows that affect sensitive people the world around. The worst, it appears, is an allergy to indoor air, especially office buildings, which affects Jane so intensely she had to quit a good job in Boston.

"I suffer from the same affliction," says Barkley.

"No you don't," says Kata.

"I do," he says. "I hate offices, I think they're poisonous."

We seem to have called a halt here. Jane has settled herself on a patch of haircap moss and is sitting with her legs stretched in front of her, daintily eating crackers and cheese from a napkin spread on her lap. Steve, the dark man from Lincoln, has his head against a tree and his arms folded on his chest. He seems half asleep. The others are leaning on their elbows, staring out to the east.

Suddenly a great thundering jet rises up.

"There's another problem around here," the carpenter points out. "That's some kind of military plane. They all collect here as if we're on the verge of war. Sometimes in June they come in force and go screaming overhead in some tedious air show business. People love it. You know, the foreign threat."

"You mean Ethiopians?" Randall says.

"Right, the Canadians might invade any day now, I heard."

"We should never have separated from the mother country," Barkley says. "They would have fought our wars for us."

This last brings Steve to life from his tree trunk. He mutters something about Punkatasset Hill, and the bridge, and all that "bullshit."

"What if there never had been a war?" he says. "What if the British won? British government outlawed the trade in human flesh not long after the Revolution, and they tried to stop the slavers at sea. Africans here and abroad would have been a lot better off if the British had stayed in power."

Barkley sits up now. "Is it not true," he asks, "that some of our most illustrious American families, mine included, for all I know, made a lot of money in the slave trade?"

Steve sits up, crosses his legs, and eyes Barkley. "Right. Never mind your high-toned Boston morality. Money's money."

"You can tell he's a professor of history," his wife says apologetically.

"It's all right," Barkley says. "We forgive him."

"Look at it this way," Steve says. "For all we know, the British might have maintained the Quebec line and prevented expansion into the American West. Tribes of indigenous people living in the region might have fared a lot better than they do today."

"It's true," Kata says. "Sitting Bull and his people used to call the north 'Mother Canada.' They were safe there."

"Isn't this all revisionist history, though?" Jane says softly. "What do they say on the television? Monday-morning halfbacks?"

A little flock of titmice brushes through the hilltop whining and chattering; Barkley glances at them perfunctorily.

"It's Monday-morning quarterbacking, I believe," he says delicately.

"It's true, though," says Diane. "You guys just talk a good line. It would have all been lost anyway. General Gage, General Washington, General Custer, what's the difference, all pigs—"

"Yeah?" says her husband, "Where would you rather be an Indian, the U.S. or Canada?"

"He's just joking," Diane says.

"As ye sow, so also shall ye reap," says Barkley.

The carpenter looks at him but says nothing.

Steve eyes the passing flock of titmice, raises his binoculars and studies them briefly, then leans back against his tree again.

Above us, the late afternoon wind makes little sallies against the desperate leaves. A few let go and drift down. A long, sustained trill of meadow crickets is coming off the hayfields just north and east of the hill, and from the surrounding shrubbery, a snowy tree cricket begins to sing. There is a tinge of melancholy in the song. It's autumn; the first hard frosts are not two weeks ahead, and soon the winter snows will close down all this pulsing life of Estabrook.

"Maybe we should go," says Randall.

By the morning of April 19, the militia and minutemen of Concord had gathered at Wright's Tavern, which still stands in the center of town. While they were there, an alarm rider named Reuben Brown came in from Lexington Green; he had seen the British volley on the common. When the assembled forces in Concord heard the news, they all agreed that, given the circumstances, the town must be defended.

One of the greatest stirrers up of strife in these hours was a minister, William Emerson, grandfather of Ralph Waldo. "Let us stand our ground," he said. "If we die, let us die here."

Concord in these years was a mere one hundred and forty years old, but the concept of *túwanasaapi* was well ingrained.

A few companies marched out of town to the east and climbed a hill to review the scene. There, in the distance, they saw their fate. A long line of regulars was snaking toward Concord, their steely muskets glinting in the morning sun, their scarlet coats and white trousers standing out against the greening fields.

On their side, the British commanders sighted the assembled Colonials on the hill and ordered a skirmish line. As the Colonials watched in awe, the single line of march split and fanned out, and the deadly machine advanced towards the hill. Inasmuch as prudence is the better part of valor, our colonial forefathers saw fit to retreat.

Back in town, the Concord companies reassembled on a hill just above the meetinghouse. Once again the British column appeared, now closer to the town, and this time the men of Concord could hear the ominous pulse of their war drums, a relentless throbbing which grew louder and louder as they slowly drew near. The vision of this mighty force, the high screaming of the fifes, the beat of the drums, struck fear into the hearts of some of the unseasoned young warriors. One of them, an eighteen-year-old named Harry Gould, was near panic, but the warlike Reverend Emerson slapped on him the back, told him to stand his ground, and know that the Lord was with him. "Our cause is just," he said.

The Concordians were at this time heavily outnumbered, and their leader, Colonel Barrett, decided that it was best to retreat once more. They marched north, crossed the North Bridge, and headed up through the Estabrook Country to regroup on Punkatasset Hill.

We are backtracking yet again, something we seem to have done repeatedly today. In a loose line we weave down through the light open oak forest of the north slope of Punkatasset Hill and then pick up the trail that runs along the southwest end of Hutchins Pond, then turn west to get back to the old Carlisle Road where the old lime quarry was located. This part of the Estabrook is the so-called Concord Field Station, maintained by Harvard's Museum of Comparative Zoology. It may be one of the more thoroughly analyzed plots of land in the whole United States. Periodically, in warmer months here, you will see bands of students with insect nets, plant presses, jars, clipboards, and other paraphernalia out on collecting forays. Today the students are still in Cambridge and the woods are given over to more mindless pleasures. I notice that on Punkatasset Hill and elsewhere away from trails we are free from the ever present hikers, dog walkers, and mountain bike devils. But as soon as we get down by the pond, we begin to encounter people again: An older gentleman with a white mustache passes us, moving slowly; he smiles and nods, as if to apologize for his sluggish progress. Two young men outfitted in outdoor catalogue clothing stride by quickly. One of them lifts his head to Kata in greeting. "How you doing?" he says. Two women in their forties pass, so engaged in conversation they barely glance up when they walk by. I catch the last phrase of a sentence, "...last I heard she was going to try regression therapy..." A man with a small white dog at his heels jogs by, drenched in sweat; there are two kids on mountain bikes, a whole extended family, toddlers riding the shoulders of the adults...all things that love the sun are out of doors today.

The carpenter is not fond of this section of the Estabrook. He's a lone woods walker and prefers to make marathon passages along the eastern boundaries of the tract each morning before anyone else appears in his woods. Sometimes, he says, even at that time he meets people—"weirdos," as he calls them. Now, in autumn, he also sees hunters who stray in from the town woods of Carlisle, where hunting is permitted.

On the way down the hill we belatedly begin the process of introduction. Randall and the carpenter are natives of Concord; their families both moved into the town in the late 1800s, Randall's from the mid-West, the carpenter's from Cambridge, where he says his family made a lot of money in the brick-making business. Randall's family may have been a part of a phenomenon that took place in Concord after the centennial celebrations of the fight at the bridge, in 1875. Many of the pilgrims who came to the town in that year, as well as those starry-eyed mid-Westerners who came to attend Bronson Alcott's School of Philosophy, were so impressed with Concord they ended up moving there. In spite of their family histories, Randall and the carpenter are not considered *true* natives. Their families are comparative latecomers.

Steve is from New York City and moved to this area to accept a teaching position; Diane is from California. She says she came here after moving from city to city and region to region. When she got to Concord, something happened.

"I'll never explain it," she says. "It was like a weight in my heels, something pulling me down into the earth. It might have been the trees. I'd never lived where there were so many old trees, but more likely it was the rivers, the rivers coming together. Anyway, something did it."

"Right," says Kata. "The Americans always used to say that the confluence of rivers is sacred territory. Spirit forces converge there, they say."

"Americans said that?" Jane asks,

"Right, Indians. So-called."

Robins suddenly burst from the woods and sail across our path.

"I had a similar experience, though, in the Berkshires before I settled there," Kata says. "I was looking for a place and I used to go to this hill and play my flute. One day I was walking near the hill and I heard a chorale of trained voices coming off the summit. They were singing a hymn, a Christian hymn. I used to play that very tune on my flute. But the thing is the words—the last verses say that if you do not praise the stones of the earth, they will sing their own praises, something like that. Anyway, I knew what it meant. The Hopi call it *túwanasaapi,* 'place where you belong.' I moved there a week later. Been there ever since. Maybe it was a church choir having a picnic in the hilltop grove. Maybe not. I never investigated."

Two doves waddle in the trail in front of us and then dash into the trees in a flurry of wings. More robins dart by, then there's a raucous chorus of blue jays, and then a general alarm among all the resident birds.

"Something's afoot," says Steve.

Even as he speaks a blue streak whips through the trees, twisting this way and that. The birds go silent, the sound of the crickets comes up in crescendo, and the woods are still.

"Cooper's hawk," says Barkley.

"Not a sharpy?" asks Randall. She means a sharp-shinned hawk.

"No, I don't think so," says Barkley. The bird contingent begins to search the trees with their binoculars to see if they can spot the attacker.

Jane is still pondering Kata's nomenclature.

"If Indians are the Americans," she asks, "what does that make us?"

"Well, where are you from?" Kata asks.

"Oh, I don't know, what do you mean, anyway? I mean, I was born in Maine, but my mother was from Indiana, and my father came from Bangor, but his family came originally from...I don't know, really. All mixed up. Dutch, I think."

"White," Kata says. "European. My Hopi family once asked me, 'Who are you?' I had to think about it all night. The next morning I came out and said, 'I am White Woman.' They liked that. A Sioux family I lived with years ago used to call me Red Willow Woman, but I like White Woman better."

"So who are we?" Jane asks.

"Where are we is more to the point," I say.

"Où allons nous?" says Barkley, without looking away from his binoculars.

Just to the north of where we stand, the Carlisle contingent was moving through the Estabrook. To the west, companies from Groton, Littleton, and Westford were now approaching the area of the bridge. Concord's Major John Buttrick and his collection of Middlesex Minutemen from Acton, Bedford, and Lincoln were in the field. All told there were now about five hundred colonials ready for a fight. Women and children had collected around the force as well, and also a number of village dogs. In those days, the hills and low grounds were clear of trees, and the assembled could see all the way to Concord, a mile distant.

By this time, the British had entered the town and were going about the business of routing out the stored arms. They found a cache of lead bullets and threw them in the millpond. They tore down the liberty pole, and found a number of wooden gun carriages, which they piled in the common. They also discovered three cannons buried in the yard of a local tavern. But for weeks, having been warned of the British raid by Paul Revere and his alarm riders, Concord residents had been moving arms out of the town or hiding them. In the last days before the arrival of the British, Colonel Barrett's sons plowed their fields as any local swain would do in spring. Then, in an act reminiscent of Jason and the Argonauts sowing dragons' teeth, the boys planted the forbidden weapons in the furrows, and covered them over. They dug them up again after the British were routed from the town.

It used to be said that Thoreau's legacy of the practice of civil disobedience—put to its greatest test by Mahatma Gandhi during the Indian struggle for independence—could only work against a civilized nation such as Britain. It is indeed possible that we all should have such chivalrous enemies. April 19 in Concord, is a case in point.

Early that morning, when Paul Revere was arrested on the road for spreading the alarm, the British officer who caught him was very angry but, being British, ever polite. He did manage to work himself up into a temper, though. "God damn you, sir," he said at the height of his rage. Then he released his prisoner.

The regulars were instructed to set the wooden gun carriages they had discovered on fire to keep them from being used again. Unfortunately, the fire spread to the nearby Town Hall, and British soldiers joined townspeople in the bucket brigade to put out the blaze. During the search and discovery, Major Pitcairn held a pistol to a tavern keeper's head to force him to reveal the hiding place of the cannons. Then he bought breakfast at the tavern for his men and insisted on paying.

The same thing happened at Colonel Barrett's house, where the spy network had informed the British that arms were hidden. After searching the place in vain, the troops requested breakfast from Mrs. Barrett. She prepared a meal and they sat to down eat—while out in the fields beyond the kitchen door the Barrett boys' deadly crop of weapons was germinating. When their meal was finished, the troops offered to pay. She refused. Filthy lucre, from filthy hands. They gave her a few shillings anyway and departed.

This, as the colonials would say, was the oppressive tyrant.

During the search of Concord, Colonel Smith, who was in command of the force, ordered six companies to hold the North Bridge. Four companies proceeded on to the Barrett house in order to search for arms, while a force of seven companies under the command of Captain Walter Laurie stayed behind to guard the bridge.

Up on Punkatasset Hill the natives were growing restless. The leaders consulted one another and decided to move the force closer to the river. The women and children were dispersed, and the soldiers attempted to chase off the local dogs who circled at their heels, sensing perhaps the onset of a hunt. Above the bridge on a flat-topped hill the militia drew up once again, and the British, seeing the presence of this larger force, withdrew down the hill towards the bridge and reassembled on the west bank. There, in the morning light, the two opposing forces faced each other. And waited.

We hike on beyond the pond and turn west. There are ancient white pines here, vast dark boles that soar for sixty feet before limbs appear. We begin to climb a slope and then dip down into a dark hollow overhung with hemlocks and high banks dotted with huge, ominous boulders. The place looks very familiar to me, but before it even registers, Barkley identifies it.

"I know this place," he says. "You do too, Kata. It's in *The Divine Comedy,* the one illustrated by Gustave Doré. We have the book."

As soon as he says the words I make the connection. I used to pore over that edition as a child, horrified and fascinated by the hordes of naked souls tormented by snakes and demons and rains of fire in Dante's version of hell. The first illustration in the book is of Dante and Virgil in the dark wood. One of the pilgrims is looking over his shoulder at the hideous leopards and lions that lurk in the shadowed trees. This hollow, with the low, narrow path running through it, and the dark overhanging trees, looks very much like the fourteenth-century Tuscan forest as envisioned by the nineteenth-century French illustrator Gustave Doré. No wonder the earliest settlers of Concord had problems with the American continent. The wild forest, as imagined across the centuries, was malevolent and threatening.

The earliest record of New England mammals, William Wood's *New England's Prospect,* doesn't do anything to disabuse readers of the myth. Those lost in the local woods were reported to have heard terrible roarings of "Devills or Lyons," ravenous, howling wolves, "strong-armed" bears, and grim-faced "ounces."

Dante, who also had a problem with the wildwood, was rescued from this secular dark forest by Virgil, the same Virgil who says in the *Aeneid* that Rome was formerly a deep forest populated by a barbarous race of beast-men who grew from hard oaks and had no skills, no settled life. They had to be overthrown in order to estab lish civilization.

Nothing new here, even in Virgil's time. In the earliest known work of literature, the Mesopotamian *Epic of Gilgamesh,* the hero leaves the sunlit walls of the city of Uruk and travels to the forests of Cedar Mountain, the domain of the gods. There he meets the hideous forest demon Humbaba, who was appointed by the Sumerian deity Enlil to protect the interests of nature against the needs of civilization. Gilgamesh enters the forest and begins chopping down the cedars. The enraged Humbaba charges, there is a pitched battle, and Gilgamesh emerges victorious. Then he cuts down all the trees. So begins Western civilization.

This attitude endured for the next four thousand years. Not until you get to Concord in 1852, right here in the Estabrook, in fact, do you get an alternative. "I went into the woods because I wished to live deliberately..."

Henry Thoreau writes in *Walden,* "to front only the essential facts of life, and see if I could not learn what it had to teach...."

By nine o'clock on the morning of April 19, the British had been on the road since ten the night before. The colonials had been up since midnight. The two groups were now at a standoff, and the tension was building.

Captain Laurie did not like the looks of the crowd on the hill above the bridge and sent a man back to Concord to call for more force.

The as yet unbanished dogs wandered between the two armies; they sat on their haunches; they lay down in the sun; they waited.

An Englishman with the colonials on the hill grew so bored with the waiting he handed over his rifle to his fellow militiamen and went down the hill to talk to his countrymen. He was down by the bridge for a long time. When he came back he took his gun and said he was going to go home.

The colonials could see the clustered spires of the churches and the Concord town house from where they stood. Suddenly, from the center of the village, they saw billowing smoke above the steeples, the result of the fire accidentally set when the British burned the gun carriages.

There was a young firebrand among the minutemen named Joseph Hosmer. In town meetings and other gatherings he was forever speaking his mind, even though he was young and worked as a journeyman. Now he launched into another one of his tirades, prodding his commander, Colonel Barrett, into action, something that would be unheard of in a professional army. "Will you let them burn the town down?" he asked.

We pass out of the hollow, climb again, cross a stream, and then ascend once more to a small hill where vast glacial boulders stand like ancient menhirs, alone in the forest. After ten more minutes we come once more to the old Carlisle Road, but instead of turning, we cross over and hike a few yards back into the untracked woods. Here there is a deep narrow cleft in the earth, overgrown with mosses and ferns, with cool, dripping granite walls where liverworts grow. In 1775 this would have been an active lime quarry where our heroes at the bridge came to mine lime. The walls of Barrett's house, of Buttrick's, Hosmer's and the other Concord men at the bridge were no doubt plastered with lime from this spot. Now it is a wild chasm. After comers could not guess the business in this place.

While their countrymen were assembled at the bridge, the men from the Westford company were still hiking along the Carlisle Road, perhaps close to this very spot. Their commander, Lieutenant Colonel John Robinson, who was on horseback, had arrived at Punkatasset Hill sometime earlier in the morning and had joined Major Buttrick among the men above the bridge. A few other Westford men had also arrived, including the Reverend Joseph Thaxter, who had been so inspired by his own sermon that he had joined the ranks. He stood next to Robinson armed with a brace of pistols. A Westford man named Joshua Parker was also there, a member of the family for whom Parker Village, which we had passed earlier, was named. So was Private Oliver Hildreth, progenitor of Hildreth Street at the base of Prospect Hill, the first street we crossed on our journey.

For a few minutes we stand staring down into the lime quarry, and then sit down again. All but Jane have logged a good twelve miles so far today and the temptation to linger in pleasant spots is setting in. Steve in particular seems fatigued. He lies down again with his head against a tree and closes his eyes.

Kata has found some beautiful green-capped mushrooms with pure white gills and is showing them to the ever curious Jane.

"Are they poisonous?" Jane asks.

"Won't kill you," Barkley says, "but I think these are incredibly peppery. You wouldn't want to eat them. They're russulas; I've forgotten the species."

"Hmmm," Jane says quietly. "The new world."

"Well they're an Old World species, too," Barkley says.

"No, I mean this is a new world to me."

"It's going to be one world soon enough," Barkley plunges on. "Half the plants around here come from Europe."

"Really?" says Jane. "How?"

"Brits," says Steve, from his bed.

"Actually, they began coming five hundred and two years ago today," Barkley points out.

"Columbus did it," says Kata.

Columbus left seeds of native European plants with the failed colony he attempted to establish at Navidad on Hispaniola in 1493, and he brought even more on the next voyage: wheat, chickpeas, onions, and sugar cane. The Spanish would commonly import herds of pigs, and they also brought along horses and dogs. Inadvertently in these first voyages, with the feed for these animals, in ballast, and mixed in with the agricultural seeds, came the first seeds of the plants we refer to as alien weeds. Ponce de León and company helped bring them over to the mainland on his second voyage to Florida. Expecting eternal life, he came equipped to stay, with farm implements, seeds, horses, and a company of men. Now we can see the results of this invasion scattered around at our feet here in the Estabrook Woods—buckthorn, and gill-over-the-ground, and beggar-ticks.

Steve seems to have fallen asleep. His wife, Diane, is staring blankly into the cleft of the lime quarry; the others are lounging again. No one seems anxious to hike these last two miles to Concord.

"After this, I'm going home to cultivate my garden," says Barkley. "I'm going to get a house with a green lawn and a white picket fence, and I'll have tended flower beds with delphiniums and peonies and a little arbor at the back of the garden with a love seat. I'll putter there with a watering can on summer mornings and watch the hollyhocks bloom, never to travel again. The soul is no traveler."

"You already have a house with delphiniums and hollyhocks," Kata says. "You never stay home long enough to enjoy them. He always talks like this when he's tired," Kata says to the group.

"No, I mean it. Traveling is a fool's paradise. It's a symptom of a deeper unsoundness."

The others stare at him silently.

"So who said that?" Kata asks.

"Me... Well, Emerson, actually."

On the hill above the bridge the assembled colonials considered the challenge of young Hosmer. Lieutenant Colonel John Robinson was older and of a higher rank than Major Buttrick, and there is a local tradition that Robinson was offered the command, but since the full company of his men had not yet arrived, he refused.

The commander of the whole group was Colonel Barrett, whose house the regulars were at that moment searching and whose food they would soon be consuming. Barrett eyed the British companies at the bridge, glanced—one presumes—at the underling upstart, Joseph Hosmer, and then ordered his men to load. He walked up and down the ranks, cautioning his soldiers. Then he gave the order to advance.

Brothers, these people from the unknown world will cut down our groves, spoil our hunting and planting grounds, and drive us and our children from the graves of our fathers....

—Metacom, AKA King Philip, 1675

Where today is the Pequot? Where are the Narragansetts, the Mohawks, the Pokanoket and many other once powerful tribes of our people? They have vanished before the avarice and the oppression of the White Man, as snow before a summer sun.

—Chief Tecumseh, 1811

The Bridge

IN 1675, ONE HUNDRED YEARS BEFORE THE EVENTS at the bridge, the true natives of the Concord region attempted to band together and oust the European invaders from their territory. Depending on whose version of history you accept, these people had been living in Concord and its environs for the last twelve to fifteen thousand years, and had been on the American continent since the beginning of time, according to their own histories, or for as many as thirty thousand years, according to the histories prepared by archaeologists of European ancestry.

The idea of driving out the invaders was started by a Pokanoket leader named Metacom, and in concept was simplicity itself. The disparate tribes would stop fighting each other and cooperate in a war effort to force the English to leave forever the American shores. It was a good idea, but it was some two hundred years too late. As Kata likes to say, if the Indians just hadn't been so *nice* to the Spanish—if they had killed them on sight instead greeting them as friends, or even gods—neither Columbus nor any subsequent expeditions would ever have returned with word of the "new" world.

For four years, Metacom, who probably had no idea that there were already Spanish settlements in Florida and the Southwest, had been brooding over the English problem. During this period he solidified his relations with his

Indian allies. His intent was to save his land, not necessarily to make war; in fact, given the Indian division of power, which had separate leaders for war and for matters of peace, it is not clear that he had total control over his people. But the members of the Pokanoket tribe must have shared his feeling: they were being pressed.

One of the first causes of the war was the fact that the English had usurped cherished agricultural lands near the Indian villages at Sowams (present-day Warren, Rhode Island) and Mount Hope. Here, in the moderate coastal environment on the protected shores of Narragansett Bay, the growing season was long, frosts came late in the autumn, and the Indian women raised excellent crops. The loss of the fields, the loss of their place, galled the local Indians.

At the root of the outbreak of the war, though, was the fundamental difference between the Indian and English concepts of land tenure. Individual allotments of land and individual ownership was an established principle of English law. But for the natives of the Americas, the land was held by the tribe, and the right to its use was a common right, like the right to the crops and to the game taken on the hunt. The principle of holding land individually and then excluding others must have been difficult for the Indians to comprehend, even though they had been dealing with the English for more than fifty years.

In 1661 the young Metacom applied for a new name from the English at Plymouth and was subsequently called King Philip, after Philip of Macedonia, father of Alexander the Great. The English should perhaps have been more careful with their names and better observers of native customs. Indians commonly took on a new name just before or during a time of crisis. It was a sign that something was about to happen.

Pleasant though it is at the lime quarry, the autumn afternoon is drawing to a close. We're out of food and drink, the easy wind that fisted above in the dry leaves has dropped, and somewhere to our west we hear the plaintive song of a white-throat sparrow.

The long sad whistle wakes Steve from his nap.

"Haven't had one of those today," he says, without identifying exactly what it is that he hasn't had. Everyone but Jane seems to understand.

"He means that birdsong," I tell her. "It's a migratory sparrow. You only see them around here in spring and fall."

She nods indifferently. She too is looking tired now.

Without discussion we get up and wander back out to the road and turn south. Just as we come to the road, a black Labrador retriever, its fur wet and caked with mud, trots past, glances over at us without breaking stride, and carries on in a businesslike manner.

"Headed home from work," Barkley says.

"Right," says Randall. "Been a hard day at the office."

Ten years after he changed his name this King Philip was under suspicion of plotting war. He had "sold"—probably under pressure—land around Swansea in southeastern

Massachusetts in 1668 and 1669 and it was known he and some of the younger sachems in the Narragansett and Pokanoket tribes were dissatisfied and stirring for trouble. In particular there were rumors of a pending attack on Swansea.

Much of the information the English had about Philip came from a Christianized Massachusetts Indian named John Sassamon who had been educated at Harvard and had worked as a sort of secretary and go-between for Philip and the English. His name appears as a witness on several of the land-sale documents that Philip was involved with, and although he may have been a man of peace—he was a preacher in the Christian church—the Indians suspected him of spying for the English.

In the winter of 1674 Sassamon disappeared. His body was later found under the ice in Assawamsett Pond near Swansea, his neck broken. Three Indians, including one of Philip's counselors, a man named Tobias, were charged with the crime by the English and, after a brief mock trial, were hanged. Philip was a suspect and was summoned to testify, but whether in defiance or fear he failed to show up. The hangings were at best a lynch party. The evidence consisted of portents: comets had been seen and blood poured forth from the mouth of the victim when the accused were paraded past the corpse.

Philip was enraged. He issued a statement at Providence some time later: the English who came first to this land were a forlorn, distressed people, he said. His father had helped them and they flourished. How did they repay this kindness? They seized his brother and confined him, he was thrown into illness and died. The Indian people were disarmed. Their land was taken by trickery, by sales.

Only a small dominion of his ancestral lands remained.

"I am determined," said King Philip, "not to live until I have no country."

Compared to other sections of our walk, the road in this section is downright boring. It rises and falls, following the contours of the land, and runs between two roughly built stone walls, constructed in all likelihood from rocks cleared sometime in the seventeenth century by the people who made this road. But it is also easy walking, and a great relief to those of us who have been picking our way through brambles and swamps for the last thirteen miles or so. The Concord crew, by contrast, has been following known trails—of which there are many in this well-preserved region—all the way from Lincoln.

On June 17, 1675, with tensions running high, some of Philip's people returned some stray horses to the English. It might have been read as a gesture of peace, an attempt at reconciliation. But not two or three days later, some of Philip's warriors burned and ransacked houses at Swansea. News of the raid spread to Plymouth. Leaders of the settlers decreed a day of fasting to encourage God to aid his people in the coming conflagration. The fields were empty on that day, the churches full, and only one boy and one old man, excused from church, stood on guard over the pastures. At some point during his watch, the boy saw a group of Pokanoket Indians emerge from the woods. Methodically they began to slit the throats of his grazing cattle, so he raised his musket and fired. One of the Indians fell dead and was dragged away by his companions.

One hundred years later at Lexington and Concord, and again at Bunker Hill, there was much banter from the colonial commanders about not firing the first shot. The command not to fire seemed to have some mystical appeal, some ring of just cause. Pokanoket shamans had been telling Philip and his warriors the same thing for several years before the onset of the war: the natives would be victorious only if the English fired the first shot. Omens for an Indian victory over the oppressors were good.

Some hours later, in retaliation, the Indians attacked and killed six Swansea citizens on their way home from church. Within days the bells were ringing throughout New England. Word of a great uprising of vast armies of Indians, of monstrous attacks and raids, was spreading through the region. Militias were raised, and England's god was called upon repeatedly to lend assistance: the heathen horde, the dreaded insurrection, which must have haunted the English sleep for generations, was finally at hand.

All up and down our pilgrimage route we have been crossing and recrossing sites of the minor incidents of this war. At Beaver Brook, below Prospect Hill in Westford, Tom Dublet, who played an instrumental role in

English and Indian diplomatic efforts, kept his fish weir and his hut. Three miles into our walk we passed just north of the Quagana Hill, where Mary Shepard was captured by a raiding party under the control of Queen Weetamoo. And here, on the old Carlisle Road, we may actually be following the original route taken by the Christianized Indians of Nashobah Plantation after they were rounded up by the piratical mercenary Captain John Moseley and were marched, roped together by the neck, down to Concord to be watched. Never mind that they were Christians—they were also *Indians.*

Not half a mile south of the old lime quarry, Randall halts and shows us a configuration of rocks overgrown with ferns and wild geraniums. Here, she tells us, was the kiln where the stone from the quarry was calcined into lime. Now it's nothing more than a rocky outcropping, and if it weren't for the efforts of local historians, no one would know the site.

All around us we can see the evidence of the second, maybe even third, growth of the ancient forest that must have characterized this route at the time when Moseley and his captives passed by here. Oaks, older birches, and white pines above an understory of blueberry surround us. The ancient land, the primordial forest so feared by the colonists, is in fact indifferent to human affairs and resurges wherever it can. In some ways it's comforting—the benign indifference of the universe, as Barkley might say.

In Swansea the English armies began to assemble to resist the uprising. They burned over a thousand acres of corn. They attacked and demolished Philip's home base at Mount Hope and rounded up hogs and cattle.

Nearby on the Sakonnet River the Sogkonate Indians, led by the great warrior chieftess Awashonks, held a war dance. Queen Weetamoo of the Pocassets, who had been married to Philip's brother, assembled her warriors. Word spread among the Indian nations. The warrior powwows began their harangues.

There were more portents of war. In Boston, a gun sounded off in the empty air. Bullets streaked across the sky. People heard drums, the tread of armies marching in the nearby woods. When they investigated they found nothing.

In Plymouth, ghostly troops of horse companies rode back and forth across the sky.

Things did not look good. Boston let out of jail those prisoners willing to fight. The sadistic pirate Captain Samuel Moseley, accompanied by hunting dogs in the manner of the Spanish conquistadors, marched south with a troop of pirates remanded from sentences of death.

On the twenty-fifth of June, there was an eclipse of the moon.

Queen Weetamoo and Awashonks held more war dances.

English towns consolidated into garrisons.

Finally in Swansea at the end of June, strikes from both sides began and the war started. Dartmouth was attacked and burned. Philip and Queen Weetamoo moved their forces north. There were raids on the outlying English houses and farms. Nipmuck Indians, under the leader Muttawump of Quabog, attacked the English town of Brookfield. Lancaster was attacked and burned. Mary Rowlandson and her children were carried off for ransom. In Groton, a group of raiders swept into town and drove off the cattle. A week later they came back and killed a man, kidnapped another, and stole the poultry and swine. Three days later a company of four hundred Indians charged into the community and burned the town. The English fled over the ridges screaming and crying while their attackers fired on them. They fled down the Great Road past the house of Mary Shepard and finally found refuge in the one town the Indians never attacked—Concord. For a hundred years afterward, on certain nights travelers and residents heard human voices screaming and crying in Groton, above the ridges over which the people escaped.

It was a brave little war, one of the first anti-imperialist efforts since the time of Virgil, and among the earliest concerted efforts by the native people of America actually to regain lost territory. In a sense it was the first American revolution. The second one, the one that started in Concord, was essentially a civil war.

Our companions have dropped a car at the end of the old Carlisle Road where the dirt road ends and the paved Estabrook Road begins. The trail widens at this point, and there are signs of horses and dogs. We see a beer can, broken sticks, candy wrappers. Even without knowing where we were we could have told we were approaching the culture of the automobile.

Our intent on this trip has been to avoid roads, and although there is a direct route along the Estabrook Road to the National Historic Park above the bridge, we decide to cut away at this point and burrow through a tangle of blackberries and cutover forest to circumvent the roads and arrive at the park through wild land.

Our fellow travelers do not envy our mission, and with many farewells and promises to meet again we part. Jane goes with them, having accepted an offer to be driven back to her car at the Middlesex School.

We watch the five of them walk off. Steve appears to have accepted the charge of the education of Jane. Randall and Diane walk together, and my friend the long-haired carpenter strolls behind the group, his thumbs hooked in the straps of his pack, as if he's carrying a heavy load. Just before they round the last bend, he lifts his arm in a slow wave, without looking back.

We are now once more forging through unpleasant conditions, but after a few hundred yards we spot a horse pasture to our left, duck under an electric fence, and follow the easternmost wall until the pasture ends at a large brushy suburban backyard. In the corner of the wall a gentleman farmer is clearing brush with loppers and stacking the cuttings along the wall.

"Just clearing out the jungle here," he says after Barkley greets him. "Turn your back and it'll jump you."

He is wearing a tattersall shirt and khakis, and a lock of snow-white hair falls across his ruddy forehead.

"Beautiful evening," he announces. "Are you lost? You're a hell of a way off the trail."

With some difficulty we explain that we have walked here from Westford through a seventeenth-century landscape.

"Jesus Christ," he says, and goes back to cutting brush.

We cross the wall once more and weave through the bushes.

This is a densely populated area, although the backyards are large and brushy, enabling us to skirt lawns and driveways. After another hundred yards we come to a wide field with a European-style stucco building on the other side. I recognize it as one of the buildings owned by the National Park Service and we brazenly cross the open field and come at last to Liberty Street, the final road before the bridge. We cross over and turn right and walk along a stone wall to a site where a historic marker now stands. It was here, in the field beyond the wall, that the provincials gathered just before the advance. We can see for the first time the winding course of the Concord River, the vegetated brushy banks and, rising between the willows, the arc of the North Bridge.

It was on this site, on Barrett's command to advance, that the drums began to roll. The fifer, Luther Blanchard of Acton, struck up a little march, "The White Cockade." Other fifers in the companies behind him took up the tune, and the assembled group of some five hundred men began down the hill toward the British, marching in double file.

They were a larger force than the regulars, and it is speculated that at this point Barrett still hoped to avoid a fight and simply make a stand in front of the town. But the fates were weaving their deadly strategies.

Down at the bridge, not a half a mile downstream from Egg Rock, where the local tribes used to gather to hold their powwows in the years before the white man came, Captain Walter Laurie watched with alarm the deadly intent of the advancing companies and prudently ordered his men to retreat across the bridge to the east bank of the river. On their way over, some of the soldiers began pulling up planks to prevent the advance of the colonials. This enraged Major John Buttrick. The two groups were close enough to call to each other by now, and Buttrick shouted at them to leave the bridge alone. It happened to be his property they had been on just a few minutes before they retreated, and Buttrick tended to think of the bridge as *his;* he used it all the time. So did Barrett. So did they all. This was *their* land, their place. Buttrick shouted in English, of course, his mother tongue, with the flat American accent. Everybody on the hill that morning thought of themselves as British—the same language, the same hymns, as Winston Churchill would remind us some one hundred and fifty years later, in time of war.

Throughout 1675 and into 1676 the native American raids on outlying English towns continued. In the early part of King Philip's War the Indians had some successes. But the fact is that by 1675, in the East the natives were already outnumbered by English, having been decimated by diseases introduced to New England even before the first colony was established at Plymouth in 1620. Furthermore, the Indians were up against a skilled fighter named Benjamin Church, who went about the war more in the fashion of a native than a British commander. He was clever at moving his forces through the forest, and what's more, he took advantage of the long-standing animosity and mistrust which existed among the native tribes. He negotiated and split them into camps and took deserters from the cause as trusted allies. There was one prize that he was after, almost more than victory in battle, and that was the head of King Philip.

In the second summer of the war, Captain Church captured and sold native warriors as well as women and children into slavery. In one particularly successful coup, he managed to capture the wife and son of Philip and probably sold them as well, to the West Indies. Finally he gained a powerful associate by winning the sachem Awashonks over to his side. By August he had gained a major victory against Philip's men, had captured and executed Muttawump, and had managed to get Queen Weetamoo on the run. At one point during her long retreat, she tried to cross a river to escape and drowned.

Time and again Church came close to catching Philip. Once he came across his still-warm campfire, and once two of Church's men saw a figure sitting on a stump near the Taunton River. They drew a bead, hesitated, thinking it was a native ally, and then realized the figure was Philip. It was too late. He dove down the bank and made his escape. He was the fox to the hound Church. The fox is wily and knows its territory. But the hound endures.

We walk back up the road from the tablet to the national park headquarters in the former Buttrick mansion, which was built by a descendant of the good major in 1911 and was acquired by the Park Service in 1963. The Service's acquisition of land on the American, or western, side of the river made it possible to restore the old road used by Buttrick and company in the eighteenth century. We bypass the building and then in the fading light are sidetracked by the formal gardens of the former estate, well tended now by the Park Service but, according to Barkley, landscaped with essentially common and boring varieties of chrysanthemums, autumn sedum, and monkshood. Laurels, rhododendrons, and privet hedges mark and quarter the grounds, and below us, large beautiful copper beeches sweep the grasses. It's a pleasant spot and we linger on the benches above the path leading down to the river, where a few late tourists wander along the old battle road. This is October, the busiest season here at the national park, not because of war, or history, but because this is leaf season and all the world turns out to see the brilliance of the forests along the Concord and Sudbury rivers.

An American family is standing at a small display with a recording that gives, in outline, the details of the battle on the morning of April 19. They appear to be from the mid-West and are casually attired in bright sweat shirts advertising beer, and a place somewhere in Wisconsin called "Joey's Fun Bar." They have two boys, aged about eight or ten, who appear to have a budding interest in American history. Again and again they push the button on the display to begin the recording that tells of the events at the bridge.

"Cut it out, Jimmy," the mother says.

Jimmy retreats and his brother advances to push the button.

"I mean it, boys," she says.

The brother halts momentarily, then pushes the button and dashes off behind Jimmy, laughing.

The father of this troop has, beneath his ample belly, a fake-leather belt pack of the type Barkley calls a twentieth-century codpiece. He unzips it and searches for something.

"No ice cream," he says to the boys.

"Why?" Jimmy asks.

"Why not?" the brother echoes.

"Push that button one more time, no ice cream."

On the eastern side of the bridge the British companies drew up in what is known as street firing position: in narrow ranks, with the front line kneeling. Across the river they could see the raggle-taggle provincial band approaching, ununiformed, clothed in homespun and leather, marching in orderly double files, casually almost, their rifles in their left hands, the long double file snaking down the hill from the flatlands above.

At the bottom of the hill the colonials turned east and began to march up to the bridge along a straight causeway, where some of the British were still straggling to cross. Once on the other side, these last mingled with the ranks who had already formed up. The colonials came on relentlessly. There was confusion in the British ranks, alarm, the colonial force was larger and better trained than the British expected, no one was happy about any of this, there were murmurs and shouts, and then, to Captain Laurie's horror, from somewhere on the British side (so it is said—this is the American version I'm telling) a shot was fired.

The first shot flew beneath the arm of our fellow traveler on this pilgrimage, Captain John Robinson of Westford. The ball whistled past without damage and struck the young fifer, Luther Blanchard, wounding him horribly. Then there were two more shots from the British side. One of them struck Isaac Davis of Acton in the heart. The ball happened to hit a brass button on his waistcoat and then pierced a major vessel, and as he fell back, a fountain of blood arched out of his chest and drenched the men beside him. These men were farmers; they were used to the blood of butchery. But this was war.

Another volley. A ball struck Private Abner Hosmer in the head. He too fell dead.

The colonials did not break rank after these first shots. They moved forward along the causeway until they were about fifty yards from the bridge, well within the muskets' killing range. It was at this point that Major Buttrick, on whose land they were still standing, turned and shouted for the men to fire. The wolves took up the cry, and all down the line there was a simultaneous shout. And then began a withering fire.

These men knew how to shoot, and in spite of their unsoldierly appearance, they remembered their orders, fired low, and took aim at the bright scarlet uniforms of the British leaders. Of the eight officers at the bridge that morning, four were hit in that first volley. So were three privates. Nine others were wounded in that single burst. The British fired back, but they were packed in tightly beyond the bridge and only the front ranks could shoot. Disorder was setting in; the smoke of black powder was a common problem in eighteenth-century warfare, and now the cloud began to obscure the action; command was breaking down on the English side; men were firing wildly, without effect.

This was a select British force at the bridge; they were world-famous for their discipline under fire, their courage, but the "weight," as it was later phrased, of the colonial riflemen was too much. The British troops broke ranks and ran for town.

The exchange had lasted less than two minutes.

We follow the old road down the hill and turn left through the willow trees along the causeway. This is low ground here, and whenever the river floods the waters rise and cover the road, leaving Daniel Chester French's 1875 statue of the embattled minuteman standing alone on an island of stone.

Out on the bridge the light is fuller and the river has taken on that evening luminosity that seems to infuse trees and rock walls so that they glow from within. A few mallard ducks are dabbling along the banks, and to the north and south, the river curves out of sight around wooded bends. Mist is rising from the American side, and a lone muskrat emerges from the tangle of bankside vegetation and cuts across the still waters of the river, leaving a perfect V in its wake.

A Japanese man with a video camera spots it and begins filming. When the muskrat submerges, he looks over at us and smiles. Barkley bows and addresses him in Japanese. The man returns his bow and greets him, laughing in surprise. Barkley prides himself on his ability to pick up languages, and even though he is not good in Japanese, he knows enough to hold brief, formal conversations. Clearly even his limited fluency has impressed the tourist. There is much bowing and joking, but when Barkley's knowledge of words runs out, the tourist switches to nearly perfect English.

He is here at a conference at one of the nearby computer companies and is on a pilgrimage of sorts of local tourist sites. He is a cosmopolitan, well-traveled fellow, with a knowledge of Western history and a fluency in several Western languages. He has lived in Brussels and New York, and spent time in Vienna and Paris, as well as Bangkok.

"Many places," he says. "Sometimes too many, I think. Airport is my second home."

Inasmuch as we are engaged here on a unique pilgrimage, we quite naturally swing the conversation around to famous pilgrimages of Japan. Years ago, when he was in his twenties, my father joined a group of Shinto pilgrims to ascend Mount Fuji, a major pilgrimage site for Shinto and Buddhist religions. I remember his descriptions of the arduous climb, the windy, foggy conditions, and the festive state of mind of the pilgrims, in spite of the hardships. I still have the stick he used on that climb, inscribed with the numbers of the stations of the ascent.

Our tourist has also ascended Fuji.

"Not the same as it was," he says. "You see beer cans, litter, many crowds. Fuji is best seen from a distance."

We ask him about the "marathon monks" of Mount Heii whose form of worship is a one-hundred-mile run around the mountain slopes.

"Oh, yes," he says. "Fanatics." He smiles broadly.

Then we ask about the great Shikoku Island pilgrimage in which the faithful circumambulate the island, stopping at shrines along the way.

"Yes, I know about that. These people... This is not Japan. Not what you think. I am sorry." He giggles weakly. "This is what I like. Place where things once happened."

"History."

"History," he says. "You Americans make so many monuments to wars. I have traveled all around this country, and Europe. Here, you come to a tourist site, what is it? Fort something or other. In Europe, the monuments, all spiritual, Chartres, Rheims. I am interested in military history." He laughs again. "I came here for this reason. I went also to the place where Agincourt was fought. I can't find anything. A farmer says: there is a stone marker there on the site. But he doesn't know where. You know Agincourt?"

"Yes, yes," says Barkley. "Of course, 'Once more unto the breach.'"

"That was Harfleur," says the tourist.

"Well, it's in *Henry the Fifth*. 'We few, we happy few...'."

"Oh yes," the tourist says, laughing, "'we happy few, we band of brothers...' Shakespeare."

"Shakespeare-*san*," says Barkley, bowing.

Mile Fifteen

But here we are; that is a great fact, and, if we tarry a little, we may come to learn that here is best.

—Ralph Waldo Emerson, "Heroism"

I have never got over my surprise that I should have been born into the most estimable place in all the world....

—Henry David Thoreau

Shambhala 

WHEN I FIRST CAME TO THIS AREA, I lived in Concord, in the center of the town, in the midst of all the activity. It was a given that on the night of April 18, sleep was impossible for those living near the green. By day the units of militia and minutemen would begin to assemble for the reenactments that are played out each year in this spot. To and fro they would march on the green and down by the bridge, and when night fell they would still be playing. Around eleven-thirty, crowds would gather at the green and the companies would strike up the music once more, the drums would rattle, the fifes would shrill out the famous tune "The World Turned Upside Down," and then, at midnight, amid much clatter and cheering, a local doctor would thunder into the center of town on his horse and proclaim the news, "The regulars are coming..." in imitation of the arrival of Doctor Samuel Prescott. Great cheering and laughter. High spirits from the assembled. Celebrants from a local grand ball, held every year on this night, would collect on the green in their formal wear. There were toasts and drumrolls and the fifes would wail, and long after the crowds had departed, from all parts of the town you could hear drums, and "The White Cockade" and "Yankee Doodle" and, over and over again, "The World Turned Upside Down," the tune that was played by the British companies as they marched to their ships after their final defeat at Yorktown.

If you live in Concord in our time, you will not be allowed to forget that this is the birthplace of a nation.

Conveniently, perhaps necessarily, we are encouraged to forget that this very site, this confluence of two rivers, was sacred to another, more established, culture, and that this, like most new nations, was founded on the ashes of another.

By the end of 1676 King Philip, hounded mercilessly by Captain Church's forces, his whereabouts revealed time and again by the treachery of his own people, made what may have been a grievous error. He crossed the Taunton River and headed for his *querencia,* Mount Hope. One does not know, at a remove of 350 years, the mind of Philip, but it is likely that unconsciously he saw the end and preferred to die in his own territory. He at least knew the peninsula well, knew all the holes and hollows and where best to make a stand. But by this time he must have been exhausted, perhaps even out of his mind. One of his counselors suggested, tentatively, that they should, perhaps, surrender. Philip killed him.

The brother of the slain man deserted and went to Church. A few days later he led Church and a company of men to Philip's camp near a swamp at the foot of Mount Hope. The English surrounded the place, intending to attack at dawn, but someone fired a shot prematurely and Philip, half-clothed, took up his weapons and made a break. Fate arranged that he run to the exact spot where the brother of the man he had killed stood with an English soldier. Philip charged, the Indian fired, piercing Philip's heart, and he fell face-down in the mud.

Church came up and looked at the body. "Naked, dirty beast," he said.

Then he cut off his head and had the corpse quartered.

In Plymouth, where it had all begun, they spiked the head of King Philip on a pike outside the town. For decades it hung there, visible to passersby. Cotton Mather used to mock the dead, gaping skull, and more than once removed the jaw of the man he called the "blasphemous leviathan."

One hundred years later, in 1775, the descendants of these hardy Englishmen, the rebels who fought at the bridge, these brave sons of liberty, still did not brook rebellion in others. They forbade local blacks to bear arms for fear they might become infected with the spirit of liberty and start an uprising of their own. And just to make certain that the local Africans understood the consequences of infractions of the laws, they left little reminders around the city of Boston. In Charlestown, on the road to the Boston ferry, there was an old rusting cage wherein lay the bones of an African slave named Mark who, perhaps disturbed by mistreatment, or by the mere fact of his slavery, had poisoned his master. Another slave who had assisted him was burned alive at the stake. The tyrannical British passed Mark's cage on April 18 on their way to Concord to collect arms to attempt to stem the colonial quest for liberty.

We hike over the bridge and stand for a few minutes at the stone marking the grave of two British soldiers who were felled by the first rebel volley. While we are reading the lines inscribed here in 1836 by the poet James Russell Lowell, a park ranger strides out from the field leading to the Old Manse and for some reason nods to Kata, who nods back politely.

Barkley greets the man and asks a question about the fallen soldiers, but the ranger is one of the enforcement officers at the park and knows nothing of history. He's a tall beefy man of about thirty, with blond hair and green eyes. I notice that he is packing a large, well-oiled forty-five and a number of rounds of ammunition and has a club and handcuffs on his belt.

"Get a lot of crime in this park?" Barkley asks in disguised innocence. This particular national park is not exactly attractive to motorcycle gangs, robbers, rapists, and murderers, as are some of the western national parks.

"You get a lot of crime everywhere these days," he says.

"But here?" Barkley persists. "I would think the bridge would be one of the safest places in America."

"Let me tell you something," the ranger says, "you put on this uniform, you're putting a target on your back. Park rangers have the highest murder rate of any enforcement officers in the nation."

At the cut in the stone wall from which the ranger had emerged we cross an open field to the Old Manse, the house of Emerson's grandfather, the minister who so encouraged the men on the ridge above the town to defend their native place against the aggressors from overseas.

Nathaniel Hawthorne moved into the Old Manse with his wife, Sophia, on their wedding day, July 9, 1842. He spent the happiest days of his life in this place, this Eden, as he called it. Nowadays the house is preserved as a historic site and is open to the public for tours. It's closed up when we pass, though, and dark inside, the black mullioned windows absorbing what light is left in the surrounding fields.

Beyond the house, across the fields, we can see the river glinting with a dull silver. Ducks are passing overhead, the robins are calling, and catbirds are whining in the grape arbor on the south side of the house. The air is cool now, and we can smell the rich, foxy odor of Concord grapes, first bred in this place in the year 1849 by Ephraim Wales Bull.

I know a short cut to Sleepy Hollow Cemetery through the backyard of an acquaintance of mine, and we walk out of the driveway of the Manse to Monument Street. Across the way is a house with a white diamond frame beside one of its doors; in the old clapboard inside the frame is a shot hole. According to legend, as the regulars retreated back to town along Monument Street, one of the soldiers fired at the house. Generations of householders proudly preserved the wound.

Somewhere around here, shortly after the skirmish at the bridge, a young provincial named Aimee White was hurrying to join the fray with an ax in his hand. The fight was over, but two wounded regulars lay in his path. Reports vary on what actually took place when White happened upon the soldiers, but when the bodies were retrieved by the British, they had been horribly mutilated—axed and scalped by White. American versions of the event suggest that the soldiers were in pain and pleading to be dispatched and White accommodated. But among the British forces word spread that the provincials were savages who knew no bounds and mutilated the dead and dying.

White lived on to become a respectable citizen of the community.

Just before we cross the road, from out of the gloaming we see an apparition—a thin man with a gaunt face, half shaven, dressed in heavy corduroys, a tweed jacket, and a red kerchief at his neck, and with a rucksack on his back. He has an odd lumbering gait, like an old, wounded elephant. I recognize him immediately. It is a man named John, one of the regulars in these parts, who commonly walks wherever he goes and appears often on the Great Road between Concord and Littleton. John is one of a number of people who, for various unknown reasons, follow that particular route almost daily. As a regular there myself, I often see them: a Guatemalan couple who ply the road in all seasons, the woman walking a few steps behind the man; an exercise fanatic who runs determinedly in all weathers and is known locally as Tight Man, partly because there is not an ounce of fat upon him and partly because his whole demeanor is wound tightly, like a Jack Russell terrier's. There is a woman, perhaps mad, of about forty years of age, who has a lazy easy-going gait and often smiles to herself as she walks. There is another John, a balding man with bad teeth who rides an old fat-tired bicycle, much fitted with equipment—lights, baskets, panniers, and a Texas license plate that reads JESUS affixed to the seat. If you fall into conversation with John, he will tell you that today is his birthday—no matter what day it is.

This John at the bridge is not a communicative type. He used to live in an old ramshackle boarding house in Littleton that took in wayward souls, no questions asked, for very low rent—one of the few places left in the area that harbored those down on their luck. Now it's closed.

John favors costumes, some of which are inappropriate. I last saw him on the hottest day of the summer dressed entirely in heavy tweeds, with a vest, a wool shirt over a turtleneck, a scarf, a Tyrolean hat pulled down over his forehead, and a heavy knapsack—one of his signatures—strapped on his back. Now he approaches us from the north, ambling tiredly. As he passes, I nod and greet him by name.

He draws up abruptly and stares at me.

"How do you know my name?" he asks.

"We've met," I tell him. "In Littleton. We've talked."

He looks the three of us over, and hikes on without another word.

"Nice guy," Kata says.

In those communities that develop over time and in response to geography, such people will appear. Their presence is a sign of an old, rooted settlement, a sign of human failings, one of the signatures of a genuine place, for better or worse. In the official lists of the ten best places to live in America—in edge cities, in communities that are planned by developers, in neighborhoods that are gated or controlled by private police, that are uniform by design and choice—such people do not appear, and if they do, they are erased.

There is a high Victorian house on the east side of Monument Street with a courtyard and apartment in back and a sharp hill beyond it. We cut through the yard and, just as dusk falls, after thrashing through a tangled wood, come out close to Pine Ridge. In the burial ground, under the tall, shading trees the light is cooler and more somber. We cross through a flat maintenence area where brush and piles of leaves have been raked together, climb another small hill, and then descend through the nineteenth-century tombstones to a narrow, paved drive. Ahead of us is a sharp hill, known to geologists as a moulin, created by draining pockets in the ice sheet during the retreat of the glacier some fifteen thousand years ago.

Hills of this sort are characterized by dry sandy soils and were often favored as burial sites in northern communities. The one ahead of us is perhaps the most famous in all New England. Often called Author's Ridge, here lie buried Thoreau, Hawthorne, the Alcotts, Emerson, and the other stars of the Concord galaxy. Except for Père-Lachaise cemetery in Paris and Westminster Abbey in London, no other burial site in the world contains so many authors interred in so small a space. Here lie the bones of the place that is Concord, that which distinguishes this piece of ground from the Alamo, from Lexington, Gettysburg, or Yorktown. The nineteenth century created Concord as America's metaphor for itself. Now the place lives on in spite of itself.

Sarah Orne Jewett, who understood as well as any other writer the meaning of *querencia* and *túwanasaapi,* says that the ghosts of those who lived before invest a place with their auras, that the spirit lingers. More than the bridge, or any of the tourist attractions—more than Walden Pond itself—this site, these inscribed stones, this quiet earth, sustain the meaning and spirit of the place that is Concord.

The path up the hill is well worn, and like so many pilgrims before us, bent under the weight of our packs, heads lowered, walking in single file, we slowly ascend.

"Look at this poor insignificant stone," says Kata, once we reach the top. She is staring at a small square marker on the left side of the Thoreau family plot; it says simply HENRY, with his dates below. Someone has laid a bouquet of goldenrods and asters at the base of the stone. They are long since faded.

A flock of crows begins to gather in the white pine forest just to the north. We can hear cars passing on Bedford Street to our south. Someone starts up a leaf blower or a lawnmower over towards the town. The machine sputters and dies. The crows fall silent. A wind brushes through the upper leaves.

Near the Thoreau family plot is another unremarkable stone with the name Hawthorne cut into it. Nathaniel is buried there somewhere among the other members of his family. Ten steps more and another family plot, that of the Alcotts, equally obscure, identified only by their initials, as if all were equals in this venture, which in a sense they were. And then next to them, a vast, rough, rose quartz boulder, with a bronze plaque set into it. Beneath this stone lies the parent of the spiritual center that is Concord, Ralph Waldo Emerson, "the passive master," as his epitaph describes him, who was but an instrument of the Oversoul that shaped this mystic place at the confluence of two rivers.

Père-Lachaise is a veritable city of the dead, with paved streets, mausoleums, crypts, monuments, and tombs. In Westminster Abbey, great statuesque sepulchers and cenotaphs with brave inscriptions and epitaphs characterize the burial sites. But here in this dismal light are but a few rough stones with chiseled surnames. Elsewhere in Sleepy Hollow the tombstones of the good burghers boast of their earthly achievements. But the inventors of the place seem to prefer obscurity.

The light is fading faster now, and we descend the hill and walk southwest through the low ground below the surrounding glacial ridges. The great clumps of ancient rhododendrons squat like gnomes against the slopes. In the half-light we can see the white forms of mushrooms standing quietly among the graves, and then, looming suddenly on our right, we see another one of Concord's monuments glowing in the shadows. It is Daniel Chester French's Melvin Memorial, a shrine to the three Melvin brothers who died in the Civil War, erected by the surviving brother. The dark rhododendrons and pines around the white marble statue absorb the ambient light so that the figure of the half-hooded woman seems to be lit from the interior, a great still, ghostly presence. Kata and Barkley, who eschew patriotic fervor, stand for a moment in spite of themselves, speechless for once, although it could be that they are simply tired and unmoved.

All around us now the white-throats are calling, and in the sky above us, big flocks of Canada geese, restless now in this season of migration, are jockeying over the trees of the cemetery to settle in the marshes of the Great Meadows National Wildlife Refuge, which is on the river just over the hill to our north. In the stillness we can hear a distant gabbling, which rises and falls on the air as new squadrons sail in from disparate parts to join the flocks already settled on the river marshes for the night.

We weave southward through the white tombs and the overarching trees. With every breath of wind, a cascade of dry leaves descends.

Six months earlier, just north of the Everglades, at Lake Okeechobee, we had stopped at a lake camp to look for wood storks. All that day we had driven through grassy plains, marshes intercut with hardwood hammocks, flatlands of pines, sugar-cane fields, and small, unhappy towns where Jamaican cane cutters and down-at-the-mouth Mexican migrant pickers eked out a meager existence. That evening at sundown on the lake, a group of loud teenagers in low-slung cars settled beside us to consume beer. Barkley was so engrossed in a field guide on local frogs, that he failed to take notice of them.

"Leave him alone," one of them said for our benefit. "He just reading his Bible."

On the Kissimee prairie, where we stopped to look for burrowing owls and caracaras, we ate catfish and hush-puppies at a local fishing camp. A Seminole woman with a come-hither look tried to get me to stay on. "We could do some fishing," she said.

As we approached the end of our pilgrimage we began stopping less often. The journey began to take on some of the dreaded characteristics of the classic American road trip; towns flashing by, undiscovered and unexplored; back roads not taken; likely bird-watching spots passed. Only the evenings and the mornings evoked place; the rest was blur, punctuated by the haunting melodies of Kata's flute.

We came in time to the Myakka River, east of Sarasota, and camped in a state park among big Winnebagos and trailers: many retirees from points north: Wisconsin, Michigan, the Dakotas. They hung little house signs out in front of their trailers, as if to fix themselves in space, even though they were three thousand miles from home— "The Smiths'," "The Jones'," "The Rogers'." Their favorite subject was weather, mainly the current temperature that day in their home place: zero degrees Fahrenheit in Kalamazoo, snow in Bismarck, a storm watch in Racine.

We made the Gulf Coast the next day and stayed on Siesta Key, just south of Sarasota. The next morning we read in the local paper that one of those old men who assiduously search the beaches of America with metal detectors had actually found something on the Siesta Key beach. Archaeologists identified it as a sixteenth-century Spanish cross.

Somewhere south of this spot, near Port Charlotte or Fort Myers, Ponce de León made a landfall. He ascended one of the rivers until the water grew too shallow, and finding no spring, descended again and sailed north. He passed mangrove archipelagos, islands, white sand beaches. He saw the great ungainly pelicans cascading into the green waters. Also big rivers—the Peace, the Myakka—and the bay at Tampa.

We drove north and stayed with a friend of Barkley's near St. Petersburg. Itinerant salesmen with a trunk full of tools—probably stolen—appeared in the suburban neighborhood that evening and tried to sell Barkley a set of wrenches for his Mercedes. We pushed on the next day and came to a campground where hound dogs got on the trail of something and yowled nightlong. In the morning I met a sprightly old man who had heard Kata playing the flute. He said he was once director of a children's' symphonic camp outside of Cincinnati. He too used to play the flute, he said, "but now I am ninety-one years of age and out of breath."

I was becoming indifferent. It was all fable anyway, just one of the mythic places of the Western mind: Ponce de León's Fountain of Youth; Ultimate Thule, Golconda; the Emerald City; the Seven Cities of Cibola; El Dorado; Shambhala, the spiritual country beyond the known world green visions of paradise, a fabulous *querencia* where you will learn the seasons, the local flowers, the cast of light at evening. But when you end up there you dream of the snowy weather back home. Dorothy had the right idea: we shouldn't look any farther than our own backyards.

At the Tampa airport I once saw the coffins of dead Floridian pilgrims, lined up rank upon rank, to be returned to their home burial grounds in the north.

Traditional Chinese make pilgrimages to the grave sites in ancestral lands where long-dead relatives lie buried in their native earth. After the war, Czechoslovak Jews returned to the site of ancestral villages obliterated by the Nazis. Dracula, it is said, had to sleep in a coffin in his native soil.

It is perhaps fitting that there is a funeral home located between the newer Sleepy Hollow Cemetery and Concord's Old Hill Burying Ground, where many of the first settlers of the town lie buried beneath now obscured slate stones and where the American commanders at the fight at the bridge, Colonel Barrett and Major Buttrick, are interred. We cut through the parking lot of the funeral home, climb over the hill through the burying ground, and begin to descend to the town green. At the top of the hill, standing alone, is a stone marking the burial site of John Jack, a slave who died in 1773, a few months after gaining his freedom, when, as the inscription reads, "Death the grand tyrant, gave him his final emancipation...."

Ahead of us is the place where it all began: the mill dam; Wright Tavern, where the minutemen gathered in the early morning of the nineteenth; and the Meeting House of the first parish in Concord. Behind us are the Alcotts' Orchard House; the Wayside, where Hawthorne lived; and Grapevine Cottage, the home of Ephraim Bull, who developed the Concord grape. Catty-corner to us, set back among its guardian trees, is the home of Ralph Waldo Emerson.

The legendary pond is over there somewhere to the south, enveloped in darkness, quiet now, reposing, obscured. In our time night bestows its last vestiges of wilderness.

Concord is disputed territory—as perhaps it should be if it is to work as the American metaphor. One half of the community seeks to self-destruct, to subsume itself to the American dream and become an indistinguishable site of condominiums, high-rises, office parks, trailer camps, fast-food outlets, and monotonous, comfortable suburban manors—a place that is a part of the modern American vernacular landscape, indistinct, interchangeable with any other place, unidentifiable, obscure, and, above all, safe, without adventure or identity. The other half, often assisted by people from outside of the community, seeks to preserve the mythic Concord, the literary and historic landscape where the American version of *querencia* and *túwanasaapi* was invented and defined.

The most recent manifestation of this internecine conflict was the fight to save Walden Woods, which began after local and New York developers sought to construct housing units, apartments, and a high-rise office park within sight of Walden Pond. Town officials approved the plan, arguing that they could find no written law to stop the building. A few of the local preservationists, some of whom were Thoreauvian scholars with no taste for politics, fought back, thus creating a small of amount of news copy that, because this was Concord and Walden Pond and not just anyplace, managed to gain notice outside the region. This in turn began to attract media people, intellectuals, musicians, and Thoreau adherents from around the nation and, after two years and some five million dollars' worth of land purchases, the pond and its environs were saved. The developers got what they needed—money—and the preservationists got what they needed, a metaphor.

But the fight goes on. Inroads are made into the Estabrook Woods and the defenders rise up, gathering more forces as the threat increases. Plans are announced for a tourist center in Heywood's Meadow, the last green space in the center of town, and the defenders rise up. It's an old story in this country, the story of small-town America attempting to save itself from America.

At one point in the long litany of shopping malls that characterizes the northwest coast of Florida, I fell asleep in the backseat. It was raining. Kata was playing her flute, long intertwining Baroque passages that haunted the landscape. Barkley drove, seemingly mesmerized by the melodies. After some time I woke up and saw green hills, cows, a curious, humped landscape, like Wisconsin or Vermont.

"Where is this place?" I asked.

"We're taking a detour," Kata said. "I couldn't stand it anymore, the malls. Now look."

"I didn't know there was a single hill in all of Florida."

"We didn't either."

Barkley and Kata began stopping again to check things out. We made detours to places with the word "springs" in their name: Tarpon Springs, Fanning Springs, Homossa Springs. We stopped to watch fish jump on the

Suwanee River. We stopped at Crystal River, then Manatee Springs, where we camped. A cold rain. Dogs barking again.

There is no mention in the histories, but one wonders whether later in life this Ponce de León had lost his faith. Why else would he seek eternal life here on earth. He was old now, at least for the age in which he was living—fifty-eight or so. He'd fought the Caribs of Dominica and Guadeloupe. They had killed his dog Bercerrico and he never did manage to conquer them. The Spaniards of Puerto Rico had moved on to Mexico, seeking more gold. The Bahamas were depopulated, with no more slaves to be captured. The natives of Florida attempted to kill anyone who landed there. His wife, his daughters, his parrots, his slaves, were comfortably ensconced in the Casa Blanca at San Juan. He had lemons and oranges in his groves, and at night the little *coqui* tree frogs kept him awake. Birdsong at dawn. Long-winged, evil-looking frigate birds circled the harbors, tropic birds coasted the sea beach below him. He must have dreamed. He had lost his riches. Woodcuts of the man, of which there are few, and most of those no doubt apocryphal, show a gaunt conquistador with a pointed beard, a hooked nose, hollowed eyes. He must have dreamed. The lure of healing waters, of springs, of ancient Western magic. Atavistic reveries of Arcadia.

He brought priests on his second voyage to Florida, and agricultural tools: an intention, a commitment to eternity in that florid place. He and and his two caravels and his two hundred men coasted north. They ascended rivers; they fought off the inevitable attacks from the local Calusas. Coasted north, ascended rivers. Fought Indians.

Sought springs. He drank. Perhaps waited after each draught. Was his skin smoother? His sword arm stronger?

Ever northward. The Suwanee River. Fanning Springs. Tarpon Springs. Manatee Springs. Crystal River, where the local Indians had constructed vast sacred mounds to heathen deities. And then, finally, at the head of the Gulf, streaming out of the mangroves, circled by crying birds, the clear waters from the great spring the Indians called Wakulla.

We climb down the hill of the Old Burying Ground and cross the green, passing en route a sign that tells us that Jethro's Oak grew near this site. It was beneath this tree, on September 12, 1635, that Major Simon Willard and his associates bought the six square miles of land for the plantation to be known as Concord. No mention here of Egg Rock, of the meeting of waters, of the convergence of two rivers. No mention of the choice of names, of Concordia.

Somewhere around here the Westford company finally caught up with their fellow minutemen. They joined the loosely organized group of companies that spent the rest of that day harassing the regulars as they made their way back to Boston. Sometime after they left, the fourteen-year-old boy from Westford appeared in town carrying the three dozen doughnuts prepared by the women in Colonel John Robinson's family. By that time, though, the fighting men of Westford were out by Meriam's Corner shooting British. The green was emptied of soldiers; only the ashes from the recent fire remained, and the milling crowd of locals who collected to care for the wounded and recount the events of the dreadful morning.

At the north end of the green, the Colonial Inn, our final objective, is well lit and festive on this holiday evening. Little groups of tourists, local people, and Boston sojourners are sitting around the tavern tables, some are lingering on the porch, and the bar in the back of the hotel is full, with people smoking and laughing, some loud and half drunk. In one taproom, which in the past was reserved for men and still fails to draw women, there is a vast television screen, on which newscasters mouth words about a sporting event. This seems to excite two or three of the avid male watchers: "All right," they say, swiping the air with their fists, "all right."

Something has happened.

There are small parlorlike sitting rooms, there is a tearoom, several formal dining rooms, and a room in the front with a piano. This is our Tabard—or our Canterbury: the end or the beginning. We take hot rum toddies—in the traditional manner of hard-pressed travelers—and sit in the front room. One of my familiars passes, a man whose job is to fight to save Concord from itself. I hail him, introduce him, and tell him about our pilgrimage. Within minutes he and Barkley are engaged in a discussion of the Tao.

"Just when you think you've found the center," my friend says of the Tao, "it moves."

It was the local students of mysticism who rose up to protect the Walden Pond environs from New York developers. Dante Alighieri would have approved. So would the Buddha. So would Krishna. Right action.

At Wakulla Springs there is an old inn in the Spanish style, with Moorish archways, iron-grille doors, and rafters painted with Toltec and Aztec designs. We were wet, tired from camping on the hard ground, and hungry for something hearty when we arrived, and so, after we saw broiled local shrimp and navy-bean soup on the menu, we decided to check in. The halls were empty, and for a while we wandered around the lobby and adjacent rooms looking for people. Finally, at the far end of a polished wood floor, a black woman and an old white man emerged from a room and proceeded towards us slowly, speaking together as they approached. He was a bird man, come to see the wild turkeys that inhabit the grounds, and the rare limpkins that feed in the ponds along the river. The black woman patiently checked us in.

The spring is a whole river, clear and blue, which boils up out of the depths of the earth to make a run to the sea. The place is now a park, managed by the state and the University of Florida, and part of the spring has a swimming section. There was a fence and a dock with a boat that would, we were told, take us on the "Wakulla River Jungle Cruise." An old black man on the dock said the best way to see limpkins was to take the boat.

"Let's do it," said Barkley.

Kata merely shook her head cynically and got on board. She does not like packaged tours.

The black man began to tell us about the limpkins, the hard-shell turtles, and the alligators and the Wakulla Spring. Years ago in the depth of the spring they found mastodon bones, he said. This was once a famous site for fake jungle films. Tarzan movies were made here, with Johnny Weissmuller. The terrifying 1957 movie *Creature from the Black Lagoon* was filmed in this spot. Local Indians called the place "Strange Water." Now it's a tourist attraction, with jungle river trips, a chain-link fence, and a protected swimming area to keep the many alligators at bay.

The river was narrow below the spring, overhung with limbs in places. Anhingas perched in the trees along the banks; there were turtles, herons, limpkins crying in the open ponds; and ominous gator heads sank beneath the waters as we passed. At one point, we floated by an open stretch of high ground where wild turkeys were feeding.

Local historians say that in the autumn of 1521, Ponce de León rowed a tender up these waters and came to the great welling spring. By this time, word was out that one of the strange, armored beings who periodically appeared off the coast of the region was at hand, and some of the local people probably watched from the vegetation as the vessel slowly made its way upstream, the banners of the House of León drooping in the still air, the heavily armored conquistadors standing, crossbows ready.

At the river's source, the boat halted, riding in the upwelling currents. The *hidalgos* eyed the surrounds. They dipped their gourds and drank. Ponce de León waited.

And then, according to the stories, he ordered the tender back downstream.

Perhaps the Indians watched while the hollow-eyed leader drank, perhaps they singled him out as chief. And then, somewhere along these banks, as the Spanish passed, the Indians let fly with their deadly, three-pronged arrows.

There was a whirring in the thickets, the histories say, a soldier gripped his throat, spun, and flopped into the waters. Another streaming arrow, another fallen soldier, and then a full outpouring from the mangroves, arrows hissing through the air, a countershower of crossbow bolts, more arrows, and shadowy, darting forms moving along the banks, and again and again from the thickets the rain of arrows.

Juan Ponce of León was in the bow during this exchange, his sword drawn, his cloak flowing; perhaps he was energized, finally, by that which he loved the most: not life but a good fight. And yet he was mortal withal. One of the three-pronged arrows streaked out from the undergrowth and drove itself deep into the joint of his armor. He crumpled in the gunnels. His soldiers, his rowers, his crossbowmen withdrew downstream and rowed out to the caravels riding at anchor offshore.

The wound was in the thigh; the histories do not say exactly where, but it was, perhaps, high up, where the cuisses meet the tuille—which is to say in the groin area—the same place where the Fisher King of the Grail Castle suffers his eternal wound, causing a weakening, an impotence. The injury and the subsequent illness infects the surrounding countryside, the Grail legends say, lays waste a once bountiful land.

Some days later Juan Ponce of León grew feverish,

and the little armada sailed back down the coast to Cuba, the green wall of the as yet unconquered Floridian peninsula passing day after day, white birds circling, gyres of wood storks above the onshore cumulus, the strafes of pelicans, the cry of gulls. They passed Dry Tortuga, struck for the open sea, and finally reached Cuba. They did their best to make Juan Ponce de León comfortable in a casa at Havana. But after some weeks, at the age of sixty-one, in spite of the sacred waters of Wakulla, after receiving the extreme unction from his priests, Ponce de León, conquistador, succumbed.

For decades afterward no Spanish came ashore on the Florida coasts.

Boots and all we are ushered into one of the formal dining rooms of the Colonial Inn. It's all dark wood with old posters and framed antique notices and English food, as perhaps is only fitting for a *colonial* inn. We take another drink, order an appetizer of potted cheese and smoked ham made from a 1792 American recipe, select a wine, and tuck in to a hearty meal. Kata orders poached salmon with a tomato saffron sauce. Barkley has New York sirloin with a sauce of pinot noir butter and onion compote. I take the New England stew of mussels, shrimp, scallops, and squid. We eat; we toast our finely crafted pilgrimage, we linger with the remains of the wine bottle, and we begin to think about dessert. Kata orders a chocolate truffle and picks at it while we eye our fellow diners and order coffee.

An older man with a handlebar mustache and blue-white eyes sits at one table with a trim businesswoman. There is a Japanese family, a table of Concord ladies in tweeds and carefully managed hair, some business types in pinstriped suits, and a lonely man in jeans and a white shirt, who is also observing his fellow diners. He drops his eyes and pretends to be busy with his food when he spots us looking at him.

It's late now, near ten o'clock. The dining room is emptying slowly. We take a brandy in one of the taprooms, where a small traditional jazz ensemble is pouring out old, sad melodies—"Embraceable You," "Smoke Gets in Your Eyes," and "Goody, Goody." The crowd consists mostly of jovial men and women in their sixties who grew up with the music. One couple is dancing a small fox trot in the hallway, and I catch snatches of conversation.

"This is the hottest place in Concord. But at least the streets are safe."

"I love the old stuff," someone else says. "Seventeen-sixty, this same room, some guy lifted a glass of ale."

As we watch the action, our shuttle driver arrives. We embrace like long lost friends and are congratulated on our grand venture, and, arms draped around one another, we walk out to the street and the car.

On the porch, a man in German loden with a long-haired dachshund tucked under his arm brushes past us. A man in shirtsleeves and suspenders stands on the sidewalk, smoking leisurely. There are small packs of tourists, and by the obelisk monument to the veterans of the Civil War, a middle-aged couple in proper business clothes lean towards each other intimately; a liaison, Kata speculates.

The monument on the green was erected in 1867

with stones taken from the North Bridge, the idea being to link events of April 19, 1775, with those of April 19, 1861, when the Civil War began. An inscription on a bronze tablet on the east side announces that those who are honored herewith found in Concord "a birthplace home or grave."

Henry Thoreau puts it another way.

"We walked in so pure and bright a light...." he writes as he approaches Concord at the end of his essay "Walking." "The west side of every wood and rising ground gleamed like the boundary of Elysium....

So we saunter toward the Holy Land...."

On the trip home, quite abruptly at the Littleton line the lights of the commercial buildings of the Great Road give way to a darkened, rural landscape. The car moves on, tires humming, leaves scattering at the side of the road; a dog walker appears in the lights, then a lone teenager, and at one of the stables, a white horse breaks from a fence and spirits off into the blackness. All around us the fields roll back to the unhoused woods and swamps through which we have so recently trekked.

It is said that medieval pilgrims on their way to Santiago de Compostela, after so many weeks and months of travel, sometimes were unable to stop themselves from walking and continued on to Finisterre, "land's end," the westernmost point of Europe. In the Middle Ages the spot marked the edge of the known world.

Epilogue

THAT WINTER KATA WENT BACK TO SECOND MESA and spent the season with her Hopi family. On her previous visit, during a solitary walk on the mesa she had picked up a small rock, which she put in her pack and carried home to the Berkshires, without telling anyone about it. When the Hopis called to invite her back, they asked her if she would please return the rock she'd taken on her last visit.

In March Barkley bought a new set of hiking boots and went for a walk in the Atlas Mountains. He happened to arrive during Ramadan and found cold, empty hotels that would miraculously come alive after sundown. He hiked, searched for birds, and ate with the celebrants after the fast was broken after sundown each evening. A family in one of the villages was preparing to make the pilgrimage to Mecca and had collected clothes from friends, which they promised to wash in the sacred waters of the Zamzam. The owners of the clothes would put them away in a drawer and use them again only after their deaths, for burial. In this manner, the safe journey to paradise would be assured.

For my part, I stayed home and explored the undiscovered country of my own backyard.

About the Author 

JOHN HANSON MITCHELL is the author of three books that deal with the natural and human history of a single square mile of land in eastern Massachusetts. *Ceremonial Time* is a fifteen-thousand-year history of the square mile tract. In *A Field Guide to Your Own Back Yard* Mitchell used his immediate surroundings to explore the natural world. In 1987 he constructed a small cabin on the ridge above the tract and spent a year living without running water or electricity in order to understand more deeply the meaning of this particular piece of earth. The result was *Living At the End of Time,* an account of this single year in this singular spot.

A graduate of Columbia University in comparative literature, John Mitchell also attended the Sorbonne and the University of Madrid. Although long interested in natural history, he originally specialized in medieval studies. Mitchell is editor of *Sanctuary* magazine, published by the Massachusetts Audubon Society. He has lectured frequently on the importance of a sense of place. In 1994 he won the John Burroughs Essay Award for a story on the Concord River and its watershed.

Prospect Hill
Westford
Lyberty Way
Acton Street
Route 495
Nonset Brook
Butter Brook
Nashoba Brook
Main Street
Great Road
Nashoba Brook
Acton
N
E
W